José Lara Ruiz

MANUAL OF IBERIAN ORCHID POLLINATORS

José Lara Ruiz

MANUAL OF IBERIAN ORCHID POLLINATORS

Orchid pollinators

ScienciaScripts

Imprint

Cover image: www.ingimage.com

This book is a translation from the original published under ISBN 978-613-9-41123-8.

Publisher:
Sciencia Scripts
is a trademark of
Dodo Books Indian Ocean Ltd. and OmniScriptum S.R.L publishing group

120 High Road, East Finchley, London, N2 9ED, United Kingdom
Str. Armeneasca 28/1, office 1, Chisinau MD-2012, Republic of Moldova, Europe
Printed at: see last page
ISBN: 978-620-8-23800-1

Tables of contents

Abstract: This preliminary list presents the species of pollinators of the Iberian Orchidaceae.

Key words: Orchidaceae, pollinators, preliminary list, Iberian Peninsula.

Abstract: This preliminary list presents the species of the polliators of the Orchidaceae of the Iberia peninsula.

Key words: Orchidaceae, pollinators, preliminary checklist, iberian peninsula.

INTRODUCTION

Currently, the Orchidaceae family has 30,517 species, 1,451 hybrids and 1,666 accepted infraspecific taxa (World Plants: Orchid List).

The ways in which orchids attract pollinators are
can be grouped into 3 groups: 1) the offer of food (nectar) (Nilsson, 1983), 2) visual deceptions (species that pretend to be nectariferous) (Scaccabarozzi et al., 2018) and 3) sexual deceptions (Peakall, 1990). To these must be added, the search for nesting sites in the case of species of the genus *Serapias* (Vereckeen et al., 2012). This is valid for Euro-mediterranean orchids.

Pollinators were included in the database for a given orchid species if they had been observed removing or depositing pollen or if the vector carried identifiable pollinaria in a body position that is consistent with the mechanics of pollinator removal and deposition, a circumstance that applies primarily to species pollinated by euglossine bees (Dressler, 1976).

The aim of this work is to compile a preliminary list of Iberian orchid pollinators as a basis for further and more complete work.

BIOGEOGRAPHIC DISTRIBUTION OF ORCHIDS

(Pacific Islands and Antarctica have been excluded, due to their low number of species)

RegionNo. of species estimated	
Europa	592
Africa	2855
Temperate Asia	2020
Tropical Asia	9419
Australasia	1582
North America	209
South America and Mexico	14700

Sources: POWO and North American Orchid Conservation Center minus the Hawaiian and non-indigenous species (https://goorchids.northamericanorchidcenter.org

MATERIAL AND METHODS

The pollen loads of the specimens in the author's private collection and the inventories of more than 40 years of field work have been studied for this work.

RESULTS

LIST OF IBERIAN ORCHIDACEAE SPECIES

Anacamptis collina (Banks & Sol. ex Russel) R.M. Bateman, Pridgeon & M.W. Chase 1997. Chase 1997
Pollinated by long-tongued bees. Attraction mechanism: deception.
Long-tongued bees: **Apidae**: Apis mellifera, Eucera longicornis

Anacamptis coriophora R.M. Bateman, Pridgeon & M.W. Chase 1997 subsp. *coriophora*
Pollinated by long-tongued bees (Apidae, Megahilidae), short-tongued bees (Halictidae), wasps (Spheidae, Vespidae), flies (Syrphidae, Stratiomyiidae, Tachinidae), moths (Zygaenidae), beetles (Cantharidae, Cerambycidae) and field bugs. Nectariferous plant.
Long-tongued bees: **Apidae**: Apis mellifera, Bombus humilis, Bombus pascuorum, Bombus pratorum, Bombus subterraneus, Ceratina cucurbitina, Eucera nigrescens, Nomada concolor, Xylocopa iris, Xylocopa valga; **Megachilidae**: Anthidium manicatum, Megachile lagopoda, Osmia aurulenta
Short-tongued bees: **Halictidae:** Halictus quadricinctus, Sphecodes gibbus; **Andrenidae**: Andrena morio.
Lygaeus equestris
Wasps: **Sphecidae**: Ammophila campestris; **Vespidae**: Polistes dominula, Vespula vulgaris.
Flies: **Syrphidae**: Volucella bombylans; **Stratiomyidae**: Stratiomys longicornis; **Tachinidae:** Tachina fera
Moths: **Zygaenidae**: Zygaena filipendulae
Beetles: **Cantharidae**: Cantharis rustica; **Cerambycidae**: Dinoptera collaris; **Oedemeridae**: Oedemera nobilis.
Field bugs: **Lygaeidae**: Lygaeus equestris

Anacamptis coriophora subsp. *fragans* (Polloini) R.M. Bateman, Pridgeon & M.W. Chase 1997 **Pollinated by long-tongued bees (Apidae, Megachilidae), short-tongued bees (Halictidae), wasps (Scoliidae, Vespidae), moths (Zygaenidae), beetles (Cantharidae, Oedemeridae, Scarabeidae Cetoniinae) and field bugs (Lygaenidae).**
Long-tongued bees: **Apidae**: Apis mellifera, Bombus lucorum, Bombus pratorum, Nomada concolor, Xylocopa iris, Xylocopa valga, Xylocopa violacea; **Megachilidae**: Osmia bicornis, Osmia cornuta.
Short-tongued bees: **Halictidae**: Halictus simplex.
Wasps: **Scoliidae**: Megascolia bidens; **Vespidae**: Vespula vulgaris. Moths: **Zygaenidae**: Zygaena filipendulae
Beetles: **Cantharidae**: Ragonycha nigricollis; **Oedemeridae**: Oedemera nobilis; **Scarabaeidae (Cetoniinae):** Tropinota hirta.
Field bugs: **Lygaeidae**: Lygaeus equestris

Anacamptis laxiflora (Lam.) R.M. Bateman, Pridgeon & M.W. Chase 1997

Pollinated by long-tongued bees (Apidae).

Long-tongued bees: **Apidae**: Bombus hortorum, Bombus pascuorum, Bombus terrestris

Ancamptis longicornu subsp. *linkiana* F. M. Vázquez, 2015
Pollinated by long-tongued bees (Apidae).
Long-tongued bees: **Apidae**: Bombus lapidarius, Bombus pascuorum, Bombus terrestris

Anacamptis morio subsp. *champagneuxii* (Barnéoud) H. Kretzschmar & al., 2007 **Pollinated by long-tongued bees (Apidae, Megachilidae) and short-tongued bees (Andrenidae, Halictidae).**
Long-tongued bees: **Apidae**: Anthophora retusa, Apis mellifera, Bombus bohemicus, Bombus confusus, Bombus hortorum, Bombus humilis, Bombus lapidarius, Bombus lucorum, Bombus muscorum, Bombus pascuorum, Bombus pratorum, Bombus ruderatus, Bombus rupestris, Bombus subterraneus, Bombus sylvarum, Bombus sylvestris, Bombus terrestris, Bombus vestalis, Eucera hungarica, Eucera longicornis, Eucera nigrescens, Eucera nigrilabris, Eucera pollinosa, Eucera seminuda, Nomada sexfasciata; **Megachilidae:** Megachile parietina, Megachile pyrenaica, Osmia aurulenta, Osmia bicornis.
Short-tongued bees: **Andrenidae**: Andrena albopunctata, Andrena morio, Andrena ovatula; **Halictidae:** Halictus cochlearitarsis, Halictus patellatus, Halictus quadricinctus, Lasioglossum calceatum, Lasioglossum xanthophus, Seladonia tumulorum

Ancamptis morio (L.) R.M. Bateman, Pridgeon & M.W. Chase 1997 subp. *morio*
Pollinated by long-tongued bees (Apidae, Megachilidae, Melittidae), short-tongued bees (Andrenidae, Halictidae), butterflies (Pieridae), moths (Noctuidae, Sphingidae Zygaenidae) and beetles (Scarabaeidae).
Long-tongued bees: **Apidae**: Anthophora retusa, Apis mellifera, Bombus bohemicus, Bombus confusus, Bombus hortorum, Bombus humilis, Bombus lapidarius, Bombus lucorum, Bombus muscorum, Bombus pascuorum, Bombus pratorum, Bombus ruderatus, Bombus rupestris, Bombus subterraneus, Bombus sylvarum, Bombus sylvestris, Bombus terrestris, Bombus vestalis, Eucera hungarica, Eucera longicornis, Eucera nigrescens, Eucera nigrilabris, Eucera pollinosa, Eucera seminuda, Nomada sexfasciata; **Megachilidae:** Osmia aurulenta, Osmia bicornis; **Melittidae**: Dasypoda hirtipes.
Short-tongued bees: **Andrenidae**: Andrena albopunctata, Andrena ovatula; **Halictidae:** Halictus cochlearitarsis, Halictus patellatus, Halictus quadricinctus, Lasioglossum xanthophus
Butterflies: **Pieridae**: Anthocharis euphenoides, Pontia daplidice.
Moths: **Noctuidae**: Cucullia santolinae, Hadena luteocincta, Hadena plebeja; **Sphingidae**: Macroglossum stellatarum; **Zygaenidae:** Zygaena lonicerae Flies: Syrphidae: Episyrphus balteatus, Sphaerophoria scripta Beetles: **Scarabaeidae:** Tropinota hirta

Ancamptis palustris (Jacq.) R.M. Bateman, Pridgeon & M.W. Chase 1997 **Pollinated by long-tongued bees (Apidae) and moths (Zygaenidae).** Long-tongued bees: **Apidae**: Bombus muscorum, Bombus terrestris Poilillas: **Zygaenidae**: Zygaena filipendulae.

Anacamptis papilionacea (L.) R.M. Bateman, Pridgeon & M.W. Chase 1997
Pollinated by long-tongued bees (Apidae, Megachilidae) and flies (Bombyliidae). Long-tongued bees: **Apidae**: Anthophora crinipes, Anthophora retusa, Apis mellifera, Bombus humilis, Eucera barbiventris, Eucera brachycera, Eucera caspica, Eucera hungarica, Eucera longicornis, Eucera nigrescens, Nomada agrestis, Nomada basalis, Nomada integra, Nomada nobilis, Nomada panzeri; **Megachilidae:** Megachile parietina, Osmia aurulenta, Osmia

caerulescens

Flies: **Bombyliidae**: Bombylius major

Anacamptis pyramidalis (L.) Rich. 1817

Pollinated by long-tongued bees (Apidae), butterflies (Hesperiidae, Lycaenidae, Nymphalidae, Pieridae), moths (Arctiidae, Geometridae, Noctuidae, Spningidae, Zygaenidae), flies (Syrphidae) and beetles (Scarabaeidae).

Long-tongued bees: **Apidae**: Apis mellifera, Bombus hortorum, Bombus pascuorum

Butterflies: **Hesperiidae**: Erynnis tages, Ochlodes sylvanus, Spialia sertorius, Thymelicus sylvestris; **Lycaenidae**: Glaucopsyche alexis, Lycaena phlaea, Plebejus argus, Polyomnaus bellargus, Pyrgus carthami, Pyrgus malvoides, Pyrgus sidae; **Nymphalidae**: Aglais urticae, Aphantopus hyperanthus, Argynnis adippe, Argynnis aglaja, Brenthis hecate, Brenthis ino, Coenonympha glycerion, Euphrydryas aurinia, Maniola jurtina, Melanargia galathea, Melitaea athalia, Melitaea cinxia, Melitaea didyma, Melitaea parthenoides, Melitaea phoebe; **Pieridae**: Anthocharis cardamines, Aporia crataegi, Pieris brassicae, Pieris napi.

Moths: **Arctiidae:** Amata phegea**;** **Geometridae:** Siona lineata; **Noctuidae:** Apamea monoglypha, Cucullia calendulae, Heliothis peltigera, Mythimna sicula, Tyta luctuosa; **Sphingidae**: Hemaris fuciformis, Macroglossum stellatarum; **Zygaenidae**: Ascita geryon, Zygaena carniolica, Zygaena filipendulae, Zygaena lonicera, Zygaena loti, Zygaena purpuralis, Zygaena rhadamanthus, Zygaena trifolii, Zygaena viciae.

Flies: **Syrphidae**: Chrysotoxum bicinctum, Episyrphus balteatus, Eupeodes corollae, Sphaerophoria scripta

Beetles: **Scarabaeidae:** Trichius fasciatus, Tropinosa squalida.

Anacamptis robusta (T. Stephenson) R.M. Bateman 2003

Pollinated by long-tongued bees (Apidae).

Long-tongued bees: **Apidae**: Bombus lapidarius, Bombus pascuorum, Bombus terrestris.

Cephalanthera damasonium Druce 1906

Pollinated by long-tongued bees (Apidae, Megachilidae), short-tongued bees (Halictidae, Andrenidae) and wasps (Tenthredinidae).

Long-tongued bees: **Apidae:** Bombus lucorum; **Megachilidae:** Megachile willughbiella.

Short-tongued bees: **Halictidae:** Lasioglossum malachurum; **Andrenidae:** Andrena morio.

Wasps: **Tenthredinidae**: Dolerus germanicus.

Cephalanthera longifolia (L.) Fritsch 1888

Pollinated by short-tongued bees (Halictidae).

Short-tongued bees: **Halictidae:** Halictus rubicundus, Lasioglossum laeve, Lasioglosum politum, Lasioglossum smeathmanellum,

Cephalanthera rubra (L.) Rich. 1817

Pollinated by long-tongued bees (Apidae, Megachilidae), short-tongued bees (Halictidae) and beetles (Curculionidae).

Long-tongued bees: **Apidae:** Bombus pascuorum; **Megachilidae**: Chelostoma campanularum, Chelostoma distinctum, Chelostoma rapunculi, Megachile willughbiella, Osmia caerulescens Long-tongued bees: **Halictidae**: Dufourea dentiventris, Lasioglossum laeve

Beetles: **Curculionidae:** Miarus ursinus.

Coeloglossum viride (L.) Hartm. 1820
Pollinated by long-tongued bees (Apidae), wasps (Ichneumonidae, Tenthredinidae) and beetles (Cantharidae, Elateridae, Melyridae, Staphilinidae).
Long-tongued bees: **Apidae**: Apis mellifera
Wasps: **Ichneumonidae**: Ichneumon balteatus; **Tenthredinidae**: Tenthredopsis floricola.
Beetles: **Cantharidae**: Cantharis rustica, Rhagonycha fulva; **Elateridae**: Cidnopus pilosus; **Melyridae**: Malachaius bipustulatus; **Staphylinidae**: Anthophagus alpinus

Corallorhiza trifida Châtel, 1760
Pollinated by flies (Syrphidae, Empididae, Scatophagidae), short-tongued bees (Andrenidae, Halictidae) and wasps (Ichneumonidae).
Flies: **Syrphidae:** Eristalis pertinax, Syrphus ribesii**;** **Empididae:** Empis tessellata; **Scatophagidae**: Scatophaga stercoraria.
Short-tongued bees: **Andrenidae:** Andrena nigroaenea; **Halictidae:** Lasioglossum calceatum
Wasps: **Ichneumonidae**: Pimpla sodalis

Cypripedium calceolus L. 1753
Pollinated by short-tongued bees (Andrenidae, Colletidae, Halictidae), long-tongued bees (Apidae), flies (Syrphidae, Muscidae), wasps (Tenthredinidae) and beetles (Staphilinidae).
Short-tongued bees: **Andrenidae**: Andrena bicolor, Andrena bucephala, Andrena chrysosceles, Andrena cineraria, Andrena flavipes, Andrena florea, Andrena fucata, Andrena haemorrhoa, Andrena helvola, Andrena lapponica, Andrena limata, Andrena nigroaenea, Andrena nitida, Andrena praecox, Andrena scotica, Andrena subopaca, Andrena taraxaci, Andrena tibialis; **Colletidae**: Colletes cunicularius; **Halictidae**: Halictus rubicundus, Lasioglossum albipes, Lasioglossum bluethgeni, Lasioglossum calceatum, Lasioglossum fratellum, Lasioglossum fulvicorne, Lasioglossum laevigatum, Lasioglossum leucozonium, Lasioglossum marginatum, Lasioglossum morio, Lasioglossum quadrinotatum, Seladonia tumulorum.
Long-tongued bees: **Apidae**: Apis mellifera, Bombus lapidarius, Nomada fabriciana Flies: **Syrphidae**: Chrysotoxum festivum, Eristalis rupium, Meliscaeva cinctella, Pipiza bimaculata, Platycheirus albimanus; **Musciadae**: Musca autumnalis
Wasps: **Tenthredinidae**: Hoplocampa crataegi Beetles: **Staphylinidae**: Anthophagus alpinus

Dactyorhiza alpestris (Pugsley) Aver. 1983
Pollinated by long-tongued bees (Apidae).
Long-tongued bees: **Apidae**: Apis mellifera, Bombus lapidarius, Bombus lucorum, Bombus pascuorum, Bombus pratorum, Bombus sichelii, Bombus soroeensis, Bombus terrestris

Dactylorhiza cantabrica H.A. Pedersen 2006
Pollinated by long-tongued bees (Apidae, Megachilidae).
Long-tongued bees: **Apidae**: Anthophora aestivalis, Anthophora plumipes, Apis mellifera, Bombus hortorum, Bombus humilis, Bombus hypnorum, Bombus lapidarius, Bombus lucorum, Bombus muscorum, Bombus pascuorum, Bombus ruderarius, Bombus ruderatus, Bombus soroeensis, Bombus sylvarum, Bombus terrestris, Eucera hungarica, Eucera nigrescens, **Megachilidae:** Osmia bicolor, Osmia bicornis.

Dactylorhiza caramulensis Verm. D. Tyteca 1989
Pollinated by long-tongued bees (Apidae), beetles (Cantharidae, Cerambycidae,

Oedemeridae) and flies (Syrphidae).

Long-tongued bees: **Apidae**: Bombus lapidarius, Bombus pascuorum, Bombus terrestris
Beetles: **Cantharidae**: Cantharis rustica; **Cerambycidae**: Alosterna tabacicolor, Anastrangalia dubia, Dinoptera collaris, Ruptela maculata, Stenurella melanura; **Oedemeridae**: Oedemera atrata, Oedemera nobilis, Oedemera podagrariae
Flies**: Syrphidae:** Episyrphus balteatus

Dactylorhiza elata subsp. *elata*
Pollinated by moths (Zygaenidae)
Moths: **Zygaenidae**: Zygaena filipendulae

Dactylorhiza elata (Poir.) Soó subsp. *sesquipedalis* (Willd.) Sóo 1962
Pollinated by long-tongued bees (Apidae), beetles (Cerambycidae), flies (Syrphidae, Empididae), moths (Zygaenidae) and butterflies (Pieridae).
Long-tongued bees: **Apidae**: Apis mellifera, Bombus lapidarius, Bombus lucorum, Bombus pascuorum, Bombus pratorum, Bombus terrestris,
Beetles: **Cerambycidae**: Leptura fulva Flies: **Syrphidae:** Eristalis arbustorum Moths: **Zygaenidae**: Zygaena filipendulae Butterflies: **Pieridae:** Aporia crataegi

Dactylorhiza ericetorum (Linton) Aver. 1982
Pollinated by long-tongued bees (Apidae), beetles (Cantharidae, Cerambycidae, Oedemeridae) and flies (Syrphidae, Empididae).
Long-tongued bees: **Apidae**: Bombus lapidarius, Bombus terrestris
Beetles: **Cantharidae**: Cantharis rustica; **Cerambycidae**: Alosterna tabacicolor, Dinoptera collaris, Ruptela maculata, Stenurella melanura; **Oedemeridae**: Oedemera nobilis, Oedemera podagrariae
Flies**: Syrphidae:** Episyrphus balteatus; **Empididae**: Empis tessellata

Dactylorhiza fuchsii (Druce) Soó 1962
Pollinated by long-tongued bees (Apidae), short-tongued bees (Andrenidae), beetles (Cantharidae, Cerambycidae, Chrysomelidae, Dascillidae, Elateridae, Oedemeridae, Scarabaeidae), flies (Syrphidae, Empididae) and butterflies (Nymphalidae) Long-tongued bees: **Apidae**: Apis mellifera, Bombus hypnorum, Bombus lapidarius, Bombus lucorum, Bombus pascuorum, Bombus pratorum, Bombus ruderatus, Bombus sylvestris, Bombus terrestris, Bombus vestalis.
Short-tongued bees: **Andrenidae**: Andrena fucata, Andrena helvola
Beetles: **Cantharidae**: Cantharis rustica; **Cerambycidae:** Alosterna tabacicolor, Anastrangalia dubia, Anastrangalia sanguinolenta, Anoplodera sexguttata, Dinoptera collaris, Grammoptera atra, Judolia sexmaculata, Leptura aethiops, Leptura maculicornis, Pachytodes cerambyciformis, Stenurella bifasciata, Stenurella melanura, Ruptela maculata; **Chrysomelidae**: Plateumaris sericea; **Dascillidae**: Dascillus cervinus; **Elateridae:** Cidnopus pilosus; **Oedemeridae**: Oedemera femorata, Oedemera podagrariae, Oedemera virescens; **Scarabaeidae (Cetoniinae):** Trichius fasciatus, Trichius zonatus.
Flies: **Syrphidae**: Eristalis arbustorum, Eristalis horticola; **Empididae**: Empis tessellata, Rhamphomyia longipes; **Stratiomyidae**: Odontomyia hydroleon
Butterflies: **Nymphalidae**: Maniola jurtina.

Dactylorhiza incarnata (L.) Soó 1962

Pollinated by long-tongued bees (Apidae), flies (Syrphidae, Empididae) and beetles (Cantharidae, Cerambycidae, Oedemeridae).
Long-tongued bees: **Apidae**: Apis mellifera, Bombus humilis, Bombus lapidarius, Bombus lucorum, Bombus pascuorum, Bombus pratorum, Bombus terrestris
Flies: **Syrphidae:** Eristalis pertinax, Volucella bombylans, Volucella pellucens; **Empididae**: Empis tessellata, Empis vitripennis
Beetles**: Cantharidae:** Cantharis rustica; **Cerambycidae:** Alosterna tabacicolor, Paracorymbia maculicornis, Pachytodes cerambyciformis, Ruptera maculata; **Oedemeridae**: Oedemera lurida, Oedemera nobilis

Dactylorhiza insularis (Sommier) O. Sánchez & Herrero 2005
Pollinated by long-tongued bees (Apidae, Megachilidae).
Long-tongued bees: **Apidae**: Anthophora aestivalis, Anthophora plumipes, Apis mellifera, Bombus barbutellus, Bombus hortorum, Bombus humilis, Bombus hypnorum, Bombus lapidarius, Bombus lucorum, Bombus muscorum, Bombus pascuorum, Bombus ruderarius, Bombus ruderatus, Bombus terrestris, Bombus vestalis, Eucera nigrescens, **Megachilidae:** Osmia bicornis

Dactylorhiza irenica F.M.Vázquez 2008
Pollinated by long-tongued bees (Apidae), beetles (Cantharidae, Cerambycidae, Oedemeridae), flies (Syrphidae).
Long-tongued bees: **Apidae:** Bombus lapidarius, Bombus terrestris
Beetles: **Cantharidae**: Cantharis rustica; **Cerambycidae**: Alosterna tabacicolor, Clytus arietis, Dinoptera collaris, Ruptela maculata, Stenurella melanura; Oedemeridae: Chrysanthia viridissima, Oedemera nobilis, Oedemera podagrariae
Flies: **Syrphidae**: Syrphus ribesii

Dactylorhiza maculata (L.) Soó 1962
Pollinated by long-tongued bees (Apidae), short-tongued bees (Halictidae), beetles (Cantharidae, Cerambycidae, Chrysomelidae, Dascillidae, Oedemeridae, Scarabaeidae), flies (Syrphidae, Anthomyiidae, Empididae, Nemestrinidae, Scarophagidae), wasps (Tenthredinidae) and butterflies (Pieridae).
Long-tongued bees: **Apidae**: Bombus barbutellus, Bombus hortorum, Bombus muscorum, Bombus pascuorum, Bombus terrestris
Short-tongued bees: **Halictidae**: Lasioglossum calceatum
Beetles: **Cantharidae**: Cantharis rustica; **Cerambycidae**: Alosterna tabacicolor, Anastrangalia dubia, Anopledura sexguttata, Clytus arietis, Dinoptera collaris, Grammoptera atra, Judolia cerambyciformis, Ruptela maculata, Stenurella melanura; **Chrysomelidae**: Plateumaris bracata; **Dascillidae**: Dascillus cervinus; **Oedemeridae**: Chrysanthia viridissima, Oedemera atrata, Oedemera nobilis, Oedemera podagrariae; **Scarabaeidae**: Trichius fasciatus.
Flies**: Syrphidae:** Eristalis arbustorum, Eristalis hortícola, Eristalis tenax**,** Volucella bombylans**; Anthoamyiidae**: Delia nigrescens, Hylemya variata; **Empididae**: Empis lívida, Empis pennipes, Empis tessellata; **Nemestrinidae**: Fallenia fasciata; **Scatophagidae**: Scatophaga stercoraria.
Wasps: **Tenthredinidae**: Tenthredopsis quadriforis Butterflies: **Pieridae**: Pieris napi

Dactylorhiza majalis (Rchb.) Rauschert 1965
Pollinated by long-tongued bees (Apidae), short-tongued bees (Halictidae), beetles

(Elateridae) and flies (Empididae).

Long-tongued bees: **Apidae**: Apis mellifera, Bombus lucorum, Bombus pascuorum, Bombus pratorum,, Bobus ruderariu Bombus sichelii, Bombus soroeensis, Bombus terrestris, Eucera longicornis, Nomada sexfasciata
Short-tongued bees: **Halictidae**: Lasioglossum leucozonium Beetles: **Elateridae**: Actenicerus siaelandicus
Flies: **Empididae**: Empis tessellata

Dactylorhiza markusii (Tineo) H. Baumann & Künkele 1981
Pollinated by long-tongued bees (Apidae, Megachilidae).
Long-tongued bees: **Apidae**: Anthophora aestivalis, Anthophora plumipes, Apis mellifera, Bombus hortorum, Bombus lapidarius, Bombus lucorum, Bombus muscorum, Bombus pascuorum, Bombus ruderarius, Bombus ruderatus, Bombus soroeensis, Bombus sylvarum, Bombus sylvestris, Bombus terrestris, Bombus vestalis, Eucera hungarica, Eucera nigrescens, **Megachilidae:** Osmia bicornis.

Dactylorhiza romana (Sebast.) Soó, 1962
Pollinated by long-tongued bees (Apidae, Megachilidae).
Long-tongued bees**: Apidae:** Anthophora aestivalis, Anthophora mucida, Anthophora plumipes, Bombus humilis, Bombus lapidarius, Bombus pascuorum, Bombus ruderatus, Bombus sylvarum, Bombus sylvestris, Bombus terrestris, Eucera hungarica, Eucera nigrescens**; Megachilidae:** Osmia bicornis

Dactylorhiza sambucina (L.) Soó 1962
Pollinated by long-tongued bees (Apidae, Megachilidae), short-tongued bees (Andrenidae, Halictidae) and butterflies (Hesperiidae, Pieridae).
Long-tongued bees: **Apidae**: Anthophora aestivalis, Anthophora mucida, Anthophora plumipes, Apis mellifera, Bombus barbutellus, Bombus bohemicus, Bombus hortorum, Bombus humilis, Bombus hypnorum, Bombus lapidarius, Bombus lucorum, Bombus muscorum, Bombus pascuorum, Bombus ruderarius, Bombus ruderatus, Bombus soroeensis, Bombus sylvarum, Bomus sylvestris, Bombus terrestris, Bombus vestalis, Eucera hungarica, Eucera nigrescens, **Megachilidae:** Osmia bicolor, Osmia bicornis.
Short-tongued bees: **Andrenidae:** Andrena nigroaenea; **Halictidae**: Lasioglossum leucozonium Butterflies: **Hesperiidae**: Pyrgus malvae; **Pieridae**: Gonepteryx rhamni

Epipactis atrorubens (Hoffm.) Besser 1809
Pollinated by long-tongued bees (Apidae) and wasps (Vespidae).
Long-tongued bees: **Apidae**: Apis mellifera, Bombus lucorum, Bombus pascuorum, Bombus pratorum, Bombus terrestris
Wasps: **Vespidae**: Delta unguicullatum, Dolichovespula saxonica, Dolichovespula sylvestris, Odynerus spinipes

Epipactis bugacensis Robatsch, 1990 **Pollinated by wasps (Vespidae)** Wasps**: Vespidae:** Vespula vulgaris

Epipactis cardina Benito & C. E. Hermos. 1998
Pollinated by long-tongued bees (Apidae) and wasps (Vespidae).
Long-tongued bees: **Apidae**: Apis mellifera, Bombus lapidarius, Bombus pascuorum,

Bombus terrestris

Wasps: **Vespidae**: Polistes dominula, Vespula vulgaris

Epipactis distans Arv.-Touv. 1872
Pollinated by long-tongued bees (Apidae) and wasps (Vespidae). Long-tongued bees: **Apidae**: Apis mellifera, Bombus terrestris Wasps: **Vespidae**: Vespula vulgaris.

Epipactis duriensis Bernardos, T. Tyteca, Revuelta & Amich 2004
Pollinated by long-tongued bees (Apidae) and wasps (Vespidae).
Long-tongued bees: **Apidae**: Apis mellifera, Bombus lapidarius, Bombus lucorum, Bombus terrestris
Wasps: **Vespidae**: Dolichovespula sylvestris, Vespula vulgaris.

Epipactis exilis P. Delforge 2004
Pollinated by long-tongued bees (Apidae) and wasps (Vespidae). Long-tongued bees: **Apidae**: Apis mellifera, Bombus terrestris Wasps: **Vespidae**: Vespula vulgaris.

Epipactis fageticola (C.E. Hermos.) Devillers.-Tersch. & Devillers 1999
Pollinated by long-tongued bees (Apidae) and wasps (Vespidae).
Long-tongued bees: **Apidae**: Apis mellifera, Bombus lapidarius, Bombus terrestris Wasps: **Vespidae**: Polistes gallicus

Epipectis helleborine (L.) Crantz 1769
Pollinated by long-tongued bees (Apidae), short-tongued bees (Halictidae), wasps (Vespidae), beetles (Cantharidae, Coccinellidae), butterflies (Pieridae), moths (Zygenidae) and ants (Formicidae).
Long-tongued bees: **Apidae**: Apis mellifera, Bombus lapidarius, Bombus lucorum Short-tongued bees: **Halictidae**: Lasioglossum calceatum
Wasps: **Vespidae**: Dolichovespula media, Dolichovespula norvegica, Dolichovespula saxonica, Dolichovespula sylvestris, Polistes bischofii, Polistes dominula, Vespa crabro, Vespula austriaca, Vespula germánica, Vespula rufa, Vespula vulgaris
Flies: **Syrphidae:** Episyrphus balteatus. Eupeodes corollae, Paragus bicolor, Paragus tibialis, Sphaerophoria scripta, Sphaerophoria rupepellii
Beetles: **Cantharidae**: Rhagonycha fulva; **Coccinellidae**: Coccinella septempunctata
Butterflies: **Pieridae:** Pieris rapae
Moths: **Zygaenidae:** Zygaena carniolica Ants: **Formicidae**: Myrmica lemasnei

Epipactis kleinii M.B. Crespo, M.R. Lowe & Piera 2001
Pollinated by wasps (Vespidae) and beetles (Oedemeridae).
Wasps: **Vespidae**: Dolichovespula saxonica, Dolichovespula sylvestris, Polistes dominulus
Beetles: **Oedemeridae**: Chrysanthia viridissima

Epipactis leptochila (Godfery) Godfery 1921
Pollinated by wasps (Vespidae) and beetles (Oedemeridae).
Wasps: **Vespidae**: Dolichovespula saxonica, Dolichovespula sylvestris, Vespula vulgaris
Beetles: **Oedemeridae**: Oedemera nobilis

Epipactis lusitanica D. Tyteca 1988
Pollinated by beetles (Cantharidae, Oedemeridae).

Beetles: **Cantharidae**: Cantharis rustica; **Oedemeridae**: Oedemera nobilis

Epipactis microphylla (Ehrhart) Sw. 1800 **Pollinated by beetles (Cantharidae)** Beetles: **Cantharidae**: Cantharis rustica.

Epipactis muelleri Godfery 1921
Pollinated by beetles (Oedemeridae).
Beetles: **Oedemeridae**: Chrysanthia viridissima

Epipactis palustris (L.) Crantz 1769
Pollinated by long-tongued bees (Apidae), short-tongued bees (Andrenidae, Colletidae, Halictidae), wasps (Vespidae, Crabronidae, Braconidae, Chrysididae, Ichneumonidae, Pompilidae, Tenthredinidae), beetles (Cantharidae, Cerambycidae, Coccinellidae, Chrysomelidae, Elateridae, Melyridae, Oedemeridae), flies (Syrphidae, Agromyzidae, Anthomyiidae, Bombyliidae, Calliphoridae, Chloropidae, Coelopidae, Empidide, Hybotidae, Muscidae, Opomyzidae, Rhinophoridae, Sarcophagidae, Scatophagidae, Sciaridae, Stratiomyidae, Tachinidae, Tephritidae, Ulidiidae), ants (Formicidae), field bugs (Miridae) and spiders (Thomisidae).
Long-tongued bees: **Apidae:** Apis mellifera, Bombus lucorum, Bombus pascuorum, Bombus pratorum
Short-tongued bees: **Andrenidae:** Andrena bicolor, Andrena fucata, Andrena haemorrhoa; **Colletidae**: Hylaeus annulatus, Hylaeus brevicornis, Hylaeus gibbus; **Halictidae**: Lasioglossum calceatum, Lasioglossum fulvicorne, Lasioglossum morio, Lasioglossum pauxillum, Lasioglossum politum, Seladonia tumulorum, Sphecodes ephippius.
Wasps: **Vespidae**: Alastor atropos, Ancistrocerus oviventris, Ancistrocerus parietinus, Dolichovespula norvegica, Dolichovespula saxonica, Dolichovespula sylvestris, Eumenes pedunculatus, Euodynerus variegatus, Polistes biglumis, Symmorphus bifasciatus, Vespula austriaca, Vespula germánica, Vespula rufa, Vespula vulgaris; **Crabronidae:** Cerceris quinquefasciata, Cerceris rybyensis, Crabro cribarius, Crossocerus elongatulus, Crossocerus podagricus, Ectemnius confinis, Ectmnius continuus, Entomognathus brevis, Gorytes quinquecinctus, Mimesa bicolor, Passaloecus insignis, Rhopalum clavipes; **Braconidae**: Chelonus carbonator; **Chrysididae**: Chrysura simplex, Holopyga inflammata; **Ichneumonidae**: Bathythrix decipiens, Cratichneumon culex, Itoplectis viduata, Pion fortipes; **Pompilidae**: Arachnospila spissa; **Tenthredinidae**: Tenthredopsis nassata.
Beetles: **Cantharidae**: Cantharis thoracica, Rhagomycha fulva; **Cerambycidae**: Pseudovania livida; **Coccinellidae:** Coccinella septempunctata, Propylea quatordecimpunctata; **Chrysomelidae**: Smaragdina xanthpasis; **Elateridae**: Adrastus pallens; **Melyridae:** Malachius bipustulatus; **Oedemeridae**: Nacerdes gracilis, Oedemera femorata, Oedemera nobilis.
Flies: **Syrphidae:** Chrysotoxum bicinctum, Dasysyrphus venustus, Episyrphus balteatus, Eristalis tenax, Eumerus strigatus, Eupeodes latifasciatus, Eupeodes lucasi, Helophilus pendulus, Lejogaster metallina, Melanostoma scalare, Neoascia meticulosa, Neoascia podagrica, Paragus haemorrhous, Paragus tibialis, Platycheirus scutatus, Scaeva pyrastri, Sphaerophoria scripta, Syritta
pipiens, Syrphus ribesii, Syrphus torvus, Syrphus vitripennis; **Agromyzidae**: Melanagromyza aenea; **Anthomyiidae**: Botanophila brunneilinea, Delia platura, Pegoplata aestiva; **Bombyliidae**: Hemipenthes maura, Hemipenthes morio; **Calliphoridae**: Lucilia illustris, Onesia sepulchralis; **Chloropidae**: Cetema elongata, Oscinella nitidigenis; **Coelopidae**: Coelopa pilipes; **Empididae**:

Empis livida, Rhamphomyia tibialis; **Hybotidae:** Hybos culiciformis; **Muscidae:** Morellia simplex, Phaonia laeta; **Opomyzidae:** Opomyza germinationis; **Rhinophoridae:** Rinophora lepida; **Sarcophagidae:** Sarcophaga dissimilis; **Scatophagidae:** Scatophaga stercoraria; **Sciaridae:** Sciara hemerobioides; **Sepsidae** Sepsis cynisea, Sepsis neocynipsea; **Stratiomyidae**: Chloromyia Formosa, Stratiomys singularior; **Tachinidae**: Loewia phaeoptera, Macquartia tenebricosa, Phryxe vulgaris; **Tephritidae**: Oxyna flavipennis; **Ulidiidae:** Herina fondescentiae.
Ants: **Formicidae**: Formica cunicularia, Formica fusca, Formica polyctena, Formica rufibarbis, Formica rufibarbis, Lasius niger, Myrmica rubra, Myrmica ruginodes
Field bugs: **Miridae**: Plagiognathus arbustorum Spiders: **Thomisidae**: Misumena vatia

Epipactis phyllanthes G.E. Sm. 1852 **Pollinated by beetles (Oedemeridae)** Beetles: **Oedemeridae:** Oedemera nobilis

Epipactis provincialis Aubenas & Robatsch 1996
Pollinated by wasps (Vespidae)
Wasps: **Vespidae:** Dolichovespula saxonica

Epipactis purpurata Sm. 1828
Pollinated by long-tongued bees (Apidae), wasps (Vespidae) and flies (Syrphidae).
Long-tongued bees: **Apidae:** Bombus pascuorum
Wasps: **Vespidae**: Dolichovespula media, Dolicovespula saxonica, Dolichovespula sylvestris, Vespula austriaca, Vespula germánica, Vespula rufa, Vespula vulgaris
Flies: **Sayrphidae:** Eristalis tenax

Epipactis rhodanensis Gévaudan & Robatsch 1994
Pollinated by flies (Syrphidae).
Flies: **Syrphidae**: Episyrphus balteatus, Eupeodes corollae, Paragus albifrons, Paragus bicolor, Paragus tibialis, Sphaerophoria rueppellii, Sphaerophoria scripta

Epipactis tremolsii Pau 1914
Pollinated by long-tongued bees (Apidae) and wasps (Vespidae, Crabronidae).
Long-tongued bees: **Apidae:** Apis mellifera, Bombus lucorum, Bombus terrestris
Wasps: **Vespidae**: Ancistrocerus parietum, Dolichovespula norvegica, Dolichovespula saxonica, Dolichovespula sylvestris, Eumenes pedunculatus, Euodynerus variegatus, Polistes biglumis, Polistes dominula, Vespula austriaca, Vespula germánica, Vespula rufa, Vespula vulgaris; **Crabronidae:** Cerceris quinquefasciata, Cerceris rybyensis, Crabro cribrarius, Crossocerus elongatulus, Crossocerus podagricus.

Epigonium aphyllum Sw. 1814
Pollinated by long-tongued bees (Apidae) and flies (Musidae).
Long-tongued bees: **Apidae:** Apis mellifera, Bombus hortorum, Bombus lucorum, Bombus pascuorum, Bombus pratorum, Bombus terrestris
Flies: **Muscidae:** Musca autumnalis

Gennaria diphylla (Link.) Parl. 1860
Pollinated by moths (Crambidae, Geometridae, Noctuidae, Tortricidae).
Moths: **Crambridae**: Cornifrons ulceratalis, Euchromius ocellea, Eudonia augusta, Eudonia

lineola; **Geometridae**: Ascotis selenaria, Costaconvexa polygrammata, Cyclophora punctaria, Eupithecia extensaria, Gymnoscelis rufifasciata; **Noctuidae**: Agrotis segetum, Ctenuplisia limbirena, Schrankia costaestrigalis; **Tortricidae**: subsequana

Goodyera repens (L.) R. Br. 1813

Pollinated by long-tongued bees (Apidae), short-tongued bees (Halictidae) and beetles (Staphilinidae).

Long-tongued bees: **Apidae:** Apis mellifera, Bombus lapidarius, Bombus lucorum, Bombus pascuorum, Bombus pratensis, Bombus terrestris

Short-tongued bees: **Halictidae:** Lasioglossum morio Beetles: **Staphilinidae:** Euphalerum kratzii

Gymnadenia austriaca (Teppener & E. Klein) P. Delforge 1998

Pollinated by long-tongued bees (Apidae), butterflies (Nymphalidae, Pieridae) and moths (Noctuidae, Sphingidae, Zygaenidae).

Long-tongued bees: **Apidae**: Apis mellifera

Butterflies: **Nymphalidae**: Aglais urticae, Erebia alberganus, Erebia medusa, Vanessa cardui; **Pieridae**: Colias phicomone

Moths: **Noctuidae**: Autographa gamma, Cerapteryx graminis; **Sphingidae**: Macroglossum stellatarum; **Zygaenidae:** Zygaena filipendulae, Zygaena purpuralis

Gymnadenia borealis (Deuce) R.M. Bateman, Pridgeon & M.W. Chase 1997

Pollinated by long-tongued bees (Apidae), butterflies (Nymphalidae, Pieridae), moths (Crambidae, Geometridae, Noctuidae, Sphingidae, Zygaenidae).

Long-tongued bees: **Apidae:** Apis mellifera, Bombus lapidarius, Bombus pascuorum, Bombus terrestris

Butterflies: **Nymphalidae**: Aglais urticae, Argynnis aglaja, Vanessa cardui; **Pieridae**: Aporia crategi. Moths: **Crambridae:** Crambus perella; **Geometridae:** Aplocera plagiata; **Noctuidae**: Autographa gamma, Autographa pulchrina; **Sphingidae**: Macroglossum stellatarum; **Zygaenidae**: Zygaena filipendulae.

Gymnadenia conopsea (L.) R. Br. 1813

Pollinated by long-tongued bees (Apidae), butterflies (Hesperiidae, Lycaenidae, Nymphalidae, Papilionidae, Pieridae), moths (Crambidae, Geometridae, Micropterigidae, Noctuidae, Sphingidae, Zygaenidae), flies (Syrpidae, Empididae, Tachinidae) and beetles (Cerambycidae).

Long-tongued bees: **Apidae:** Apis mellifera, Bombus pascuorum, Bombus terrestris

Butterflies: **Hesperiidae**: Ochlodes sylvanus, Pyrgus carthami, Tymelicus sylvestris; **Lycaenidae:** Plebejus idas, Polyommatus icarus; **Nymphalidae**: Aglais urticae, Argynis adipe, Argynnis aglaja,

Boloria euphrosyne, Euphrydryas aurinia, Vanessa cardui; **Papilionidae**: Papilio machaon; **Pieridae**: Aporia crategi.

Moths: **Crambridae:** Crambus perlella, Eudonia sudetica, Pyrausta despicata; **Geometridae:** Aplocera plagiata, Entephria caesiata, Siona lineata; **Noctuidae**: Autographa bractea, Autographa gamma, Autographa pulchrina, Mythimna conigera, Papestra biren, Phytometra viridaria; **Sphingidae**: Deilephela porcellus, Hemaris tytius, Hyles euphorbiae, Hyles galii, Hyles livornica, Macroglossum stellatarum, Sphinx ligustri, Sphinx pinastri; **Zygaenidae**: Adscita geryon, Adscita statites, Zygaena exulans, Zygaena filipendulae, Zygaena lonicerae, Zygaena loti, Zygaena trifolii Flies: **Syrphidae**: Episyrphus balteatus, Syrphus ribesii.

Beetles: **Cerambycidae**: Grammoptera atra

Gymnadenia densiflora (Wahlenb.) A. Dietr.
Pollinated by butterflies (Hesperiidae, Lycaenidae, Nymphalidae, Papilionidae, Pieridae), moths (Geometridae, Noctuidae, Sphingidae, Zygaenidae), flies (Syrpidae, Empididae, Tachinidae) and beetles (Cerambycidae).
Butterflies: **Hesperiidae**: Ochlodes sylvanus, Tnymelicus sylvestris; **Lycaenidae:** Plebejus idas, Polyommatus icarus; **Nymphalidae**: Aglais urticae, Argynis adipe, Argynnis aglaja, Boloria euphrosyne, Euphrydryas aurinia, Maniola jurtina, Melanargia galathea, Vanessa cardui; **Papilionidae**: Papilio machaon; **Pieridae**: Aporia crategi, Pieris napi.
Moths: **Geometridae:** Aplasta onoraria, Aplocera plagiata, Entephria caesiata, Siona lineata; **Noctuidae**: Autographa bractea, Autographa gamma, Autographa pulchrina, Mythimna conigera, Noctua pronuba, Papestra biren, Phytometra viridaria, Plusia festucae; **Sphingidae**: Deilephela porcellus, Hemaris tytius, Hyles euphorbiae, Hyles galii, Hyles livornica, Macroglossum stellatarum, Sphinx ligustri, Sphinx pinastri; **Zygaenidae**: Adscita geryon, Adscita statites, Zygaena exulans, Zygaena filipendulae, Zygaena lonicerae, Zygaena trifolii.
Flies: **Syrphidae**: Syrphus ribesii; **Empididae:** Empis lívida, Empis tessellata; **Tachinidae**: Siphona geniculata
Beetles: **Cerambycidae**: Paracorymba fulva

Gymnadenia conopsea var. *pyrenaica* (Philippe) Nyman, 1882
Pollinated by butterflies (Hesperiidae, Lycaenidae, Nymphalidae, Pieridae) and moths (Geometridae, Noctuidae, Sphingidae, Zygaenidae).
Butterflies: **Hesperiidae**: Ochlodes sylvanus, Thymelicus sylvestris; **Lycaenidae:** Polyommatus icarus; **Nymphalidae**: Aglais urticae, Euphrydryas aurinia, Maniola jurtina, Vanessa cardui; **Pieridae**: Aporia crategi.
Moths: **Geometridae:** Aplocera plagiata, Entephria caesiata, Siona lineata; **Micropterigidae**: Micropterix aureatella; **Noctuidae**: Autographa bractea, Autographa gamma, Autographa pulchrina, Mythimna conigera, Mythimna vitellina; **Sphingidae**: Deilephela porcellus, Hemaris tytius, Hyles euphorbiae, Hyles galii, Hyles livornica, Macroglossum stellatarum, Sphinx ligustri, Sphinx pinastri; **Zygaenidae**: Adscita geryon, Adscita statites, Zygaena filipendulae, Zygaena lonicerae, Zygaena trifolii.

Gymnadenia gabasiana (Teppener & E. Klein) Teppener & E. Klein 1998
Pollinated by long-tongued bees (Apidae), butterflies (Nymphalidae) and moths (Crambridae, Geometridae, Noctuidae, Zygaenidae) and flies (Muscidae).
Long-tongued bees: **Apidae:** Apis mellifera, Bombus pascuorum, Bombus terrestris
Butterflies: **Nymphalidae**: Erebia melampus
Moths: **Crambidae:** Crambus perlella, Diasemia reticularis; **Geometridae:** Crocota tinctaria; **Noctuidae:** Autographa gamma, Lasionycta imbecilla, Leucania comma; **Zygaenidae**: Adscita statites, Zygaena exulans, Zygaena filipendulae, Zygaena loti, Zyganea transalpina, Zygaena trifolii Flies: **Muscidae**: Thricops simplex

Gymnadenia nigra (L.) Rchb. f
Pollinated by moths (Crambidae, Noctuidae, Zygaenidae), butterflies (Hesperiidae, Nymphalidae), long-tongued bees (Apidae) and flies (Muscidae).
Long-tongued bees: **Apidae:** Apis mellifera
Moths: **Crambidae:** Crambus perlella, Diasemia reticularis; **Noctuidae:** Autographa gamma,

Chersotis ocellina, Lasionycta imbecilla, Leucania comma; **Zygaenidae**: Adscita statites, Zygaena exula, Zygaena filipendulae, Zygaena loti, Zygaena transalpina, Zugaema troolii
Butterflies: **Hesperiidae**: Pyrgus serratulae; **Nymphalidae**: Boloria euphrosyne, Melitaea athalia, Melitaea parthenoides
Flies: **Muscidae**: Thricops simplex

*Gymnadenia odoratissima (*L.) Rich. 1817
Pollinated by butterflies (Lycaenidae), moths (Crambidae, Geometridaae, Prodoxidae, Sesiidae, Zygaenidae) and field bugs (Lygaeidae).
Butterflies: **Lycaenidae:** Aricia cramera
Moths: **Crambidae:** Catoptria pinella, Crambus perlella; **Geometridae:** Elophos dilucidaria, Entephria caesiata, Gnophos obfuscata, Odezia atrata, Perizoma obsoletata**; Noctuidae:** Lasionycta imbecilla**,** Mythimna conigera**; Prodoxidae:** Lampronia fuscatella; **Sesiidae:** Chamaesphecia empiformis**; Zygaenidae**: Adscita geryon, Zygaena lonicerae.
Field bugs: **Lygaeidae**: Lygaeus equestris, Lygaeus simulans

Gymnadenia odoratissima subsp. *longicalcarata* C.. E. Hermos. & Sabando 1996
Pollinated by butterflies (Nymphalidae) and moths (Geometridae, Noctuidae, Zygaenidae).
Butterflies: **Nymphalidae:** Maniola jurtina
Moths: **Geometridae:** Elophos dilucidaria, Gnophos obfuscata, Odezia atrata, Perizoma obsoletata**; Noctuidae:** Autographa gamma, Lasionycta imbecilla**,** Mythimna conigera, Noctua pronuba**; Zygaenidae**: Adscita geryon, Zygaena lonicerae, Zygaema trofolii

Himantoglossum hircimum (L.) Spreng. 1826
Pollinated by long-tongued bees (Apidae, Megachilidae), short-tongued bees (Andrenidae, Colletidae), beetles (Elateridae, Oedemeridae, Tenebrionidae) and wasps (Vespidae).
Long-tongued bees: **Apidae**: Apis mellifera, Bombus lucorum, Bombus terrestris; **Megachilidae**: Anthidium cingulatum, Anthidium loti, Coelioxys caudatus, Megachile ericetorum, Megachile lefebvrei, Megachile maritima, Megachile parietina
Short-tongued bees: **Andrenidae**: Andrena bicolor, Andrena carbonaria, Andrena cineraria, Andrena fulva, Andrena haemorrhoa, Andrena nigroaenea, Andrena nitida, Andrena pilipes, Andrena potentillae; **Colletidae:** Colletes cunicularius, Colletes similis
Beetles: **Elateridae**: Cidnopus pilosus; **Oedemeridae**: Oedemera nobilis; **Tenebrionidae**: Heliotaurus ruficollis
Wasps: **Vespidae**: Odynerus femoratus

Himantoglossum metlesicsianum (W.P. Techner) P. Delforge 1999
Pollinated by long-tongued bees (Apidae).
Long-tongued bees: **Apidae**: Apis mellifera, Bombus hortorum, Bombus lucorum, Bombus terrestris

Himantoglossum robertianum (Loisel.) P. Delforge 1999
Pollinated by long-tongued bees (Apidae, Megachilidae).
Long-tongued bees: **Apidae**: Anthophora plumipes, Bombus hortorum, Bombus lucorum, Bombus ruderatus, Bombus subterraneus, Bombus terrestris, Eucera interrupta, Eucera longicornis, Eucera nigrescens, Eucera tricincta, Xylocopa violacea; **Megachilidae**:

Megachile sicula

Limodorum abortivum (L.) Sw. 1799

Pollinated by long-tongued bees (Apidae, Megachilidae) and short-tongued bees (Halictidae).

Long-tongued bees: **Apidae:** Anthophora dispar, Anthophora mucida, Anthophora plumipes, Bombus terrestris, Habropoda tarsata; **Megachilidae:** Anthidium manicatum, Osmia caerulescens, Rhodanthidium septemdentatum

Short-tongued bees: **Halictidae**: Lasioglossum bluetgheri, Lasioglossum marginatum, Lasioglossum xanthopus

Limodorum trabutianum Batt. 1886

Pollinated by long-tongued bees (Apidae).

Long-tongued bees: **Apidae:** Anthophora plumipes, Bombus terrestris,

Neotinea conica (Willd.) R.M. Bateman 2003

Pollinated by long-tongued bees (Apidae, Megachilidae).

Long-tongued bees: **Apidae:** Apis mellifera; **Megachilidae**: Osmia bicornis

Neotinea maculata (Desf.) Stearn 1975

Pollinated by long-tongued bees (Apidae, Megachilidae), short-tongued bees (Halictidae) and beetles (Cerambycidae, Chrysomelidae, Melyridae, Scarabaeidae).

Long-tongued bees: **Apidae:** Apis mellifera, Nomada fucata; **Megachilidae**: Osmia aurulenta, Osmia bicolor, Osmia bicornis, Osmia niveata.

Short-tongued bees: **Halictidae**: Halictus patellatus

Beetles: **Cerambycidae**: Dinoptera collaris; **Chrysomelidae**: Spermophagus sericeus; **Melyridae**: Malachius bipustulatus; **Scarabaeidae:** Oxythyrea funesta

Neotinea tridentata

Pollinated by long-tongued bees (Apidae, Megachilidae), short-tongued bees (Andrenidae, Halictidae), flies (Syrphidae, Sarcophagidae, Tachinidae) and beetles (Cerambycidae, Elateridae, Scarabaeidae).

Long-tongued bees: **Apidae**: Apis mellifera, Bombus hortorum, Bombus terrestris, Nomada fucata; **Megachilidae**: Osmia aurulenta, Osmia bicolor, Osmia bicornis, Osmia niveata

Short-tongued bees: **Andrenidae:** Andrena cineraria; **Halictidae:** Halictus patellatus, Lasioglossum calceatum

Flies: **Syrphidae**: Eristalis pertinax; **Sarcophagidae:** Sarcophaga variegata; **Tachinidae**: Tachina magnicornis

Beetles: **Cerambycidae**: Dinoptera collaris; **Elateridade**: Ctenicera pecticornis; **Scarabaeidae:**
Oxythyrea funesta, Valgus hemipterus

Neotinea ustulata (L.) R.M. Bateman, Prigdeon & M.W. Chase 1977

Pollinated by flies (Syrphidae, Sarcophagidae, Tachinidae) and beetles (Cerambycidae, Oedemeridae).

Flies: **Syrphidae**: Eristalis pertinax; **Sarcophagidae:** Sarcophaga variegata; **Tachinidae**: Tachina magnicornis

Beetles: **Cerambycidae**: Pseudovadonia lívida; **Oedemeridae**: Oedemera nobilis

Neottia cordata (L.) Rich. 1817
Pollinated by flies (Limoniidae, Mycetophilidae, Sciaridae, Tipulidae) and wasps (Braconidae, Ichneumonidae).
Flies: **Limoniidae**: Limnophila schranki; **Mycetophilidae**: Mycetophila Formosa; **Sciaridae**: Sciara flavimana; **Tipulidae:** Tipula oleracea
Wasps: **Braconidae**: Microgaster procera; **Ichneumonidae**: Aptesis femoralis, Cratichneumon culex, Cratichneumon flavifrons, Cratichneumon semirufus, Cratichneumon viator, Ichneumon insidiosus, Ichneumon ligatorius, Ichneumon submarginatus, Ichneumlon suspìciosus, Perilissus pallidus, Phygadeuoun cephalotes, Phygadeuon fumator, Phygadeuoun rugolosus, Rhorus neustriae, Tryphon bidentatus

Neottia nidus-avis (L.) Rich. 1817
Pollinated by flies (Syrphidae) and ants (Formicidae).
Flies: **Syrphidae**: Meliscaeva cinctella, Syrphus ribesii
Ants: **Formicidae**: Leptotothorax gredleri, Paratrechina longicornis

Neottia ovata (L.) Bluff & Fingerh. 1838 Long-tongued bees: **Apidae**: Apis mellifera
Short-tongued bees: **Andrenidae**: Andrena chrysosceles, Andrena haemorrhoa, Andrena nitida;
Colletidae: Colletes cunicularius
Wasps: **Argidae**: Arge cyanocrocea, Arge pagana, Arge rustica, Arge ustulata; **Cimbicidae**: Abia fasciata; **Crabronidae**: Argogorytes mystaceus; **Ichneumonidae**: Aethecerus nitidus, Agrypon flaveolatum, Allomya debellator, Aptesis assimilis, Aptesis femoralis, Campoletis zonata, Chorinaeus longicornis, Coelichneumon impressor, Cratichneumon culex, Cratichneumon flavifrons, Cratichneumon semirufus, Cratichneumon viator, Cryptus leucocheir, Ctenochira pastoralis, Diplazon laetatorius, Dusona inermis, Endasys brevis, Ethelurgus sodalis, Exetastes illusor, Glyphicnemis vagabunda, Heteropelma amictum, Hyperbatus segmentator, Ichneumon insidiosus, Ichneumon ligatorius, Ichneumon submarginatus, Ichneumlon suspìciosus, Idiolispa analis, Limerodops subsericans, Lissonota cruentator, Mesoleptus laevigatus, Netelia lineolata, Perilissus lutescens, Perilissus pallidus, Phygadeuoun cephalotes, Phygadeuon fumator, Phygadeuoun rugulosus, Pimpla melanacrias, Rhorus femoralis, Rhorus neustriae, Symphertha antilope, Townesia tenuiventris, Trychosis pauper, Tryphon bidentatus, Tryphon nigripes, Tryphon signator, Vulgichneumon suavis; **Tenthredinidae**: Aglaostigma aucupariae, Amauronematus histrio, Claremontia brevicornis, Dolerus anticus, Dolerus bajulus, Dolerus cothurnatus, Dolerus germanicus, Dolerus gonager, Dolerus niger, Dolerus nitens, Dolerus picipes, Dolerus planatus, Dolerus vestigialis, Empria immersa, Empria pumila, Empria sexpunctata, Eutomostethus ephippium, Macrophya albicincta, Macrophya duodecimpunctata, Pachyprotasis rapae, Rhogogaster viridis, Tenthredo atra, Tenthredo crassa, Tenthredo lívida, Tenthredo olivácea, Tenthredo temula, Tenthredo vespa, Tenthredopsis nassata, Tentredopsis quadriforis
Beetles: **Cantharidae**: Cantharis coronata, Cantharis franciana, Cantharis instabilis, Cantharis
nigra, Cantharis palliata, Cantharis paulinoi, Cantharis reichei, Cantharis xanthoporpa, Rhagonycha divisa, Ragonycha nigriceps; **Cerambycidae**: Alosterna tabacicolor, Clytus arietis, Dinoptera collaris, Grammoptera atra, Stenurella melanura, Stenurella nigra; **Chrysomelidae**: Bruchus atomarius, Bruchus loti; **Coccinellidae**: Coccinella septempunctata; **Curculionidae:** Curculio nucum; **Dascillidae:** Dascillus cervinus; **Dasytidae**: Dasytus plumbeus; **Elateridae**: Agriotes acuminatus, Athous nigricornis, Athous

vittatus, Cidnopus pilosus, Ctenicera pastoralis, Dalopius marginatus, Denticollis linearis; **Melyridae:** Malachius bipustulatus; **Nitulidae:** Meligethes aeneus; **Oedemeridae:** Ischnomerera cinerascens, Oedemera femorata, Oedemera nobilis, Oedemera `podagrariae, Oedemera virescens; **Orsodacnidae:** Orsodacne cerasi; **Scarabaeidae (Cetoniinae):**

Valgus hemipterus; **Scarabaeidae (Melolonthinae):** Hoplia bilineata, Hoplia coerulea; **Scraptiidae**: Anaspis frontalis
Flies: **Syrphidae**: Cheilosia bergensammi, Melanostoma mellinum, Meliscaeva cinctella, Pipiza bimaculata; **Anthomyiidae**: Hydrophoria ruralis; **Bibionidae**: Bibio marci; **Cylindrotomidae:** Cylindrotoma distinctissima; **Empididae**: Empis decorella, Empis tessellata; **Limoniidae**: Dactylolabis sexmaculata; **Sarcophagidae**: Sarcophaga africa; **Tachinidae:** Phorocera grandis, Zophomyia temula; **Tipulidae**: Tipula luna.
Ants: **Formicidae:** Camponotus herculeanus, Formica fusca, Formica lemani, Lasiur niger, Myrmica ruginodis
Grasshoppers: **Tettigoniidae**: Pholidoptera griseoaptera Spiders: **Thomosidae:** Misumena vatia

Ophrys algarvensis D. Tyteca, Benito & M. Walravens 2003
Pollinated by long-tongued bees (Apidae) and short-tongued bees (Colletidae).
Long-tongued bees: **Apidae**: Anthophora atroalba Short-tongued bees: **Colletidae**: Colletes albomaculatus

Ophrys alpujata Riech. & H. Kohlmüller 2019 **Pollinated by short-tongued bees (Andrenidae)** Short-tongued bees: **Andrenidae**: Andrena cinerea.

Ophrys apifera Huds. 1762
Pollinated by long-tongued bees (Apidae).
Long-tongued bees: **Apidae**: Eucera longicornis

Ophrys arachnitiformis Gren. & M. Philippe 1860
Pollinated by short-tongued bees (Andrenidae, Colletidae).
Short-tongued bees: **Andrenidae:** Andrena nigroaenea, Andrena senecionis, Andrena trimmerana; **Colletidae**: Colletes cunicularius

Ophrys araneola Rchb. 1831
Pollinated by short-tongued bees (Andrenidae), long-tongued bees (Megachilidae).
Short-tongued bees: **Andrenidae**: Andrena combinata, Andrena lathyri.
Long-tongued bees: **Megachilidae**: Osmia bicolor

Ophrys arnoldii P. Delforge 1999
Pollinated by short-tongued bees (Andrenidae).
Short-tongued bees: **Andrenidae**: Andrena nigroaenea

Ophrys atlantica Munby 1856
Pollinated by long-tongued bees (Megachilidae).
Long-tongued bees: **Megachilidae**: Megachile parietina

Ophrys balearica P. Delforge 1999

Pollinated by long-tongued bees (Megachilidae).
Long-tongued bees: **Megachilidae**: Megachile sicula

Ophrys battandieri E.G.Camus 1908
Pollinated by short-tongued bees (Andrenidae).
Short-tongued bees: **Andrenidae**: Andrena vetula

Ophrys bilunulata Risso 1844
Pollinated by short-tongued bees (Andrenidae).
Short-tongued bees: **Andrenidae**: Andrena flavipes, Andrena nigroaenea

Ophrys bombyliflora Link 1799
Pollinated by long-tongued bees (Apidae).
Long-tongued bees: **Apidae**: Eucera algira, Eucera gracilipes (I. Canarias), Eucera grisea, Eucera nigrescens, Eucera oraniensis

Ophrys castellana Devillers -Tersch & Devillers 1988
Pollinated by short-tongued bees (Andrenidae, Colletidae) and beetles (Oedemeridae)
Short-tongued bees: **Andrenidae:** Andrena schencki; **Colletidae**: Colletes cunicularius
Beetles**: Oedemeridae:** Oedemera podagrariae

Ophrys catalaunica O. Danesch & EW. Danesch 1972
Pollinated by long-tongued bees (Megachilidae).
Long-tongued bees: **Megachilidae:** Megachile albonotata, Megachile parietina, Megachile pyrenaica

Ophrys clara F.M. Vázquez & S. Ramon 2005 **Pollinated by short-tongued bees (Colletidae)** Short-tongued bees: **Colletidae:** Colletes cunicularius

Ophrys corbariensis (J. Samuel & J.-M Lewin) Kreutz 2005
Pollinated by long-tongued bees (Apidae).
Long-tongued bees: **Apidae:** Eucera interrupta, Eucera longicornis, Eucera nigrescens

Ophrys decembris S. Moingeon & J.M. Moingeon 2011 **Pollinated by short-tongued bees (Colletidae)** Short-tongued bees: **Colletidae:** Colletes cunicularius

Ophrys dianica M.R, Lowe, Piera, M.B, Crespo & J.E. Arnold 2001
Pollinated by short-tongued bees (Andrenidae) and long-tongued bees (Megachilidae).
Short-tongued bees: **Andrenidae**: Andrena vulpecula Long-tongued bees: **Megachilidae**: Megachile albonotata

Ophrys fabrella Paulus and Ayasse ex P. Delforge 2004 **Pollinated by short-tongued bees (Andrenidae)** Short-tongued bees: **Andrenidae**: Andrena fabrella

Ophrys ficalhoana J.A. Guim. 1887
Pollinated by long-tongued bees (Apidae).
Long-tongued bees: **Apidae**: Eucera dimidiata, Eucera nigrescens, Eucera nigrilabris

Ophrys forestieri (Rch. f.) Lojac. 1909
Pollinated by short-tongued bees (Andrenidae, Colletidae).
Short-tongued bees: **Andrenidae:** Andrena bicolor, Andrena flavipes, Andrena florea, Andrena nigroaenea; **Colletidae**: Colletes cunicularius

Ophrys funerea Viv. 1824
Pollinated by short-tongued bees (Andrenidae).
Short-tongued bees: **Andrenidae**: Andrena flavipes, Andrena ovatula, Andrena wilkella

Ophrys fusca Link 1799 subsp. *fusca*
Pollinated by short-tongued bees (Colletidae, Andrenidae) and long-tongued bees (Apidae).
Short-tongued bees: **Colletidae:** Colletes cunicularius; **Andrenidae:** Andrena flavipes, Andrena florentina, Andrena fulvicrus, Andrena haemorrhoa, Andrena nigroaenea, Andrena nigroolivacea, Andrena ovatula, Andrena similis, Andrena thoracica, Andrena trimmerana
Long-tongued bees: **Apidae**: Bombus subterraneus

Ophrys fusca subsp. *iricolor* (Desf.) K. Richt., 1890
Pollinated by short-tongued bees (Colletidae, Andrenidae) and long-tongued bees (Megachilidae).
Short-tongued bees: **Colletidae:** Colletes cunicularius; **Andrenidae:** Andrena morio Long-tongued bees: **Megachilidae:** Megachile sicula

Ophrys fusca subsp. *limensis* F. M. Vázquez 2009 **Pollinated by short-tongued bees (Colletidae)** Short-tongued bees: **Colletidae:** Colletes cunicularius

Ophrys incubacea Bianca ex Tod. 1842
Pollinated by short-tongued bees (Andrenidae) and long-tongued bees (Apidae) Short-tongued bees: **Andrenidae**: Andrena morio
Long-tongued bees: **Apidae**: Melecta albifrons

Ophrys insectifera subsp. *aymonii* Breist. 1981 **Pollinated by short-tongued bees (Andrenidae)** Short-tongued bees: **Andrenidae**: Andrena combinata

Ophrys insectifera L. 1753 subsp. *insectifera*
Pollinated by wasps (Crabronidae) and beetles (Scarabaeidae) Wasps: **Crabronidae:** Argogorytes fargeii, Argogorytes mystaceus Beetles: **Scarabaeidae**: Trichius fasciatus

Ophrys kallaikia C.E. Hermos. 2018
Pollinated by short-tongued bees (Andrenidae).
Short-tongued bees: **Andrenidae:** Andrena nitida ssp. hispaniola

Ophrys lenae M. R. Lowe & D. Tyteca 2012 **Pollinated by long-tongued bees (Apidae)** Long-tongued bees: **Apidae:** Anthophora atroalba

Ophrys lucentina P. Delforge 1999
Short-tongued bees: **Andrenidae:** Andrena vulpécula

Ophrys lupercalis Devierllers- Tersch. & Deviller, 1994 **Pollinated by short-tongued bees (Andrenidae)** Short-tongued bees: **Andrenidae**: Andrena nigroaenea

Ophrys lutea subsp. *galilaea* (H. Fleischm. & Bornm.) Soó **Pollinated by short-tongued bees (Andrenidae)** Short-tongued bees: **Andrenidae**: Andrena cinerea

Ophrys lutea Cav. 1793 subsp. *lutea*
Pollinated by short-tongued bees (Andrenidae).
Short-tongued bees: **Andrenidae**: Andrena bicolor, Andrena cinerea, Andrena humilis, Andrena nigroaenea, Andrena nigroolivacea, Andrena ovatula, Andrena senecionis, Andrena taraxaci, Andrena tibialis, Andrena trimmerana

Ophrys malacitana M.R. Lowe, I. Phillips & Paulus 2010 **Pollinated by short-tongued honeybees (Colletidae)** Short-tongued honeybees: **Colletidae:** Colletes cunicularius

Ophrys marzuola (Geniez, Melki & Soca) Soca 2020
Pollinated by short-tongued bees (Andrenidae, Colletidae).
Short-tongued bees: **Andrenidae:** Andrena nigroaenea, Andrena senecionis; **Colletidae**: Colletes cunicularius

Ophrys montserratensis Cadevall 1904
Pollinated by long-tongued bees (Megachilidae).
Long-tongued bees: **Megachilidae:** Megachile parietina

Ophrys oceanica Soca 2024
Pollinated by long-tongued bees (Apidae).
Long-tongued bees: **Apidae:** Eucera interrupta, Eucera longicornis, Eucera nigrescens

Ophrys omegaifera subsp. *dyris* (Maire) Del Prete 1984
Pollinated by long-tongued bees (Apidae, Megachilidae) and short-tongued bees (Andrenidae).
Long-tongued bees: **Apidae**: Anthophora atriceps, Anthophora atroalba, Anthophora balearica;
Megachilidae: Osmia bicornis
Short-tongued bees: **Andrenidae:** Andrena flavipes

Ophrys picta Link 1799
Pollinated by long-tongued bees (Apidae).
Long-tongued bees: **Apidae**: Eucera barbiventris, Eucera cineraria

Ophrys pintoi M. R. Lowe & D. Tyteca 2012
Long-tongued bees: **Apidae**: Anthophora atroalba

Ophrys quarteirae (Kreutz, M.R. Lowe & Wucherpf.) Devillers & Devillers - Tersch. 2013
Short-tongued bees: **Andrenidae:** Andrena cinerea

Ophrys querciphila Nicole, Hervy & Soca 2017 **Pollinated by long-tongued honeybees (Apidae)** Long-tongued honeybees**: Apidae:** Tetralonia strigata

Ophrys riojana C.E. Hermos. 1999
Pollinated by short-tongued bees (Andrenidae).
Short-tongued bees: **Andrenidae:** Andrena florea, Andrena hypopolia, Andrena limbata, Andrena nigroaenea

Ophrys santonica J.M. Mathé & Melki 1994 **Pollinated by long-tongued bees (Apidae)**
Long-tongued bees: **Apidae:** Tetralonia dentata

Ophrys scolopax subsp. *apiformis*
Pollinated by long-tongued bees (Apidae).
Long-tongued bees**: Apidae:** Eucera barbiventris

Ophrys scolopax Cav. 1793 subsp. *scolopax*
Pollinated by long-tongued bees (Apidae).
Long-tongued bees: **Apidae:** Eucera barbiventris, Eucera elongatula, Eucera interrupta, Eucera longicornis, Eucera nigrescens, Eucera nigrifacies, Eucera notata

Ophrys spectabilis (Kreutz & zelesny) Paulus 2011 **Pollinated by long-tongued bees (Apidae)** Long-tongued bees: **Apidae:** Eucera rufa

Ophrys speculum subsp. *lusitanica* O. Danesh & E. Danesch 1969
Pollinated by wasps (Crabronidae)
Wasps: **Crabronidae:** Argogorytes fargeii, Argogorytes mystaceus

Ophrys speculum Link 1799 subsp. *speculum*
Pollinated by wasps (Scoliidae)
Wasps: **Scoliidae**: Dasyscolia ciliata

Oprys sphegodes subsp. *atrata* (Rchb. f) O. Bolòs, 1950
Short-tongued bees (Andrenidae)
Short-tongued bees: **Andrenidae**: Andrena nigroaenea

Ophrys sphegodes subps. *aveyronensis* J.J. Wood 1983
Short-tongued bees (Andrenidae)
Short-tongued bees: **Andrenidae**: Andrena hattorfiana

Ophrys sphegodes subsp. *litigiosa* (E. G. Camus) Bech., 1925
Short-tongued bees (Andrenidae)
Short-tongued bees: **Andrenidae**: Andrena nigroaenea

Ophrys sphegodes subsp. *passionis* (Sennen) Sanz & Nuet, 1995
Pollinated by short-tongued bees (Andrenidae).
Short-tongued bees: **Andrenidae:** Andrena pilipes

Ophrys sphegodes Mill. 1768 subsp. *sphegodes*
Pollinated by short-tongued bees (Andrenidae).
Short-tongued bees: **Andrenidae:** Andrena barbilabris, Andrena bimaculata, Andrena

cineraria, Andrena florea, Andrena limata, Andrena nigroaenea, Andrena thoracia

Ophrys subinsectifera C.E. Hermos. & Sabando 1996
Pollinated by wasps (Argidae)
Wasps: **Argidae:** Sterictiphora gastrica

Ophrys tenthredinifera Willd. 1805
Pollinated by long-tongued bees (Apidae).
Long-tongued bees: **Apidae:** Eucera algira, Eucera clypeata, Eucera dimidiata, Eucera dimidiata, Eucera longicornis, Eucera nigrescens, Eucera nigrilabris, Eucera notata, Eucera oraniensis, Eucera rufa, Eucera vidua

Ophrys vasconica (O. Danesh & E. Danesh) P. Delforge 1991
Pollinated by short-tongued bees (Andrenidae, Colletidae).
Short-tongued bees: **Andrenidae:** Andrena flavipes, Andrena nigroaenea, Andrena pilipes; **Colletidae**: Colletes cunicularius

Ophrys vitorica Kreutz 2007
Pollinated by short-tongued bees (Andrenidae).
Short-tongued bees: **Andrenidae:** Andrena hattorfiana

Orchis anthropophora All. 1875
Pollinated by beetles (Cantharidae, Cerambycidae, Elateridae, Tenebrionidae), flies (Empididae, Syrphidae) and wasps (Argidae, Chrysididae, Tenthredinidae).
Beetles: **Cantharidae**: Cantharis rutisca; **Cerambycidae**: Grammoptera atra; **Elateridae**: Cidnopus pilosus; **Tenebrionidae**: Isomira murina
Flies: **Empididae**: Empis pennipes; **Syrphidae**: Platyheirus manicatus
Wasps: **Argidae**: Arge ustulata; **Chrysididae:** Chrysis bicolor; **Tenthredinidae**: Tenthredopsis ornata
Ants: **Formicidae**: Formica fusca

Orchis canariensis Lindl. 1835
Pollinated by long-tongued bees (Apidae).
Long-tongued bees: **Apidae**: Anthophora alluaudi, Bombus canariensis, Eucera gracilipes, Eucera lanuginosa; **Megachilidae**: Megachile canescens

Orchis italica Poir. 1798
Pollinated by long-tongued bees (Apidae, Megachilidae) and beetles (Scarabaeidae).
Long-tongued bees: **Apidae**: Apis mellifera, Bombus humilis, Eucera nigrescens; **Megachilidae:** Rhodanthidium septemdentatum.
Beetles: **Scarabaeidae:** Oxytherea funesta

Orchis lapalmensis (Leibbach & Ruedi Peter) P. Delforge 2007
Pollinated by long-tongued bees (Apidae).
Long-tongued bees: **Apidae**: Bombus canariensis

Orchis mascula subsp. *ichnusae* Corrias 1982
Pollinated by long-tongued bees (Apidae, Megachilidae).

Long-tongued bees: **Apidae:** Apis mellifera, Bombus hortorum, Bombus humilis, Bombus lapidarius, Bombus lucorum, Bombus pascuorum, Bombus pratorum, Bombus ruderarius, Bombus ruderatus, Bombus rupestris, Bombus subterraneus, Bombus sylvarum, Bombus sylvestris, Bombus terrestris, Eucera hungarica, Eucera interrupta, Eucera longicornis; **Megachilidae**: Osmia bicolor.

Orchis mascula subsp. *laxifloriformis* Rivas Goday & B. Rodr. 1946 (=O. langei)
Pollinated by long-tongued bees (Apidae, Megachilidae).
Long-tongued bees: **Apidae:** Apis mellifera, Bombus lapidarius, Bombus lucorum, Bombus pascuorum, Bombus pratorum, Bombus ruderarius, Bombus ruderatus, Bombus rupestris, Bombus subterraneus, Bombus sylvarum, Bombus sylvestris, Bombus terrestris, Eucera hungarica, Eucera interrupta, Eucera longicornis; **Megachilidae**: Osmia bicolor

Orchis mascula (L.) L. 1775 subsp. *mascula*
Pollinated by long-tongued bees (Apidae, Megachilidae) and short-tongued bees (Andrenidae, Halictidae), flies (Syrphidae, Bombyliidae, Empididae, Scatophagidae), beetles (Scarabaeidae Cetoniinae), moths (Zygaenidae) and wasps (Crabronidae).
Long-tongued bees: **Apidae:** Apis mellifera, Bombus barbutellus, Bombus bohemicus, Bombus hortorum, Bombus humilis, Bombus lapidarius, Bombus lucorum, Bombus muscorum, Bombus pascuorum, Bombus pratorum, Bombus ruderarius, Bombus ruderatus, Bombus rupestris, Bombus subterraneus, Bombus sylvarum, Bombus sylvestris, Bombus terrestris, Bombus vestalis, Eucera hungarica, Eucera interrupta, Eucera longicornis, Nomada marshamella; **Megachilidae**: Osmia bicolor
Short-tongued bees: **Andrenidae:** Andrena haemorrhoa, Andrena helvola, Andrena nigroaenea, Andrena nitida; **Halictidae**: Halictus sexcinctus
Flies: **Syrphidae**: Rhingia campestris; **Bombyliidae**: Bombylius major; **Empididae**: Empis tessellata; **Scatophagidae**: Scatophaga stercoraria
Beetles: **Scarabaeidae (Cetoniinae):** Cetonia aurata Moths: **Zygaenidae**: Zygaena transalpina
Wasps: **Crabronidae**: Argogorytes mystaceus

Orchis militaris L. 1753 subsp. *militaris*
Pollinated by long-tongued bees (Apidae, Megachilidae) and short-tongued bees (Andrenidae, Halictidae), beetles (Scarabaeidae Cetoniinae), butterflies (Pieridae), flies (Syrphidae, Muscidae, Opomyzidae, Scatophagidae) and ants (Formicidae).
Long-tongued bees: **Apidae:** Anthophora aestivalis, Apis mellifera, Bombus lapidarius, Bombus lucorum, Bombus pascuorum, Bombus pratorum, Bombus terrestris, Bombus vestalis, Nomada fabriciana, Nomada ruficornis, Nomada succincta; **Megachilidae**: Hoplitis adunca, Osmia aurulenta, Osmia bicolor, Osmia bicornis
Short-tongued bees: **Andrenidae:** Andrena cineraria, Andrena curvungula, Andrena enslinella, Andrena hattorfiana; **Halictidae**: Halictus eurygnathus, Halictus simplex, Sphecodes ferruginatus Scarab beetles: **Scarabaeidae (Cetoniinae):** Tropinota hirta
Butterflies: **Pieridae:** Pieris brassicae, Pieris napi
Flies: **Syrphidae**: Baccha elongata, Leucozona lucorum, Melanostoma mellinum, Platychierus scutatus, Rhingia campestris; **Muscidae**: Thricops semicinesreus; **Opomyzidae**: Opomyza germinationis; **Scatophagidae**: Scatophaga stercoraria
Ants: **Formicidae**: Myrmica ruginodis

Orchis olbiensis Reut. ex Gren. 1860
Pollinated by long-tongued bees (Apidae, Megachilidae).
Long-tongued bees**: Apidae:** Apis mellifera, Bombus hortorum, Bombus lapidarius, Bombus lucorum, Bombus pascuorum, Bombus pratorum, Bombus ruderarius, Bombus ruderatus, Bombus rupestris, Bombus subterraneus, Bombus sylvarum, Bombus sylvestris, Bombus terrestris, Eucera hungarica, Eucera interrupta, Eucera longicornis; **Megachilidae**: Osmia bicolor

Orchis pallens L. 1771
Pollinated by long-tongued bees (Apidae, Megachilidae).
Long-tongued bees: **Apidae:** Bombus hortorum, Bombus lapidarius, Bombus pascuorum, Bombus pratorum, Bombus sylvestris, Bombus terrestris, Eucera nigrescens, Xylocopa valga**;** **Megachilidae:** Osmia cornuta

Orchis provincialis Balv. 1806
Pollinated by long-tongued bees (Apidae).
Long-tongued bees: **Apidae:** Bombus humilis, Eucera hungarica, Eucera nigrescens

Orchis purpurea Huds. 1762 subsp. *purpurea*
Pollinated by long-tongued bees (Apidae, Megachilidae), short-tongued bees (Andrenidae, Halictidae), beetles (Scarabaeidae) and butterflies (Papilionidae) Long-tongued bees*:* **Apidae*:*** Apis mellifera, Bombus terrestris, Eucera nigrescens; **Megachilidae**: Osmia aurulenta, Osmia bicornis, Osmia caerulescens, Osmia rufohirta, Osmia viridana
Short-tongued bees: **Andrenidae:** Andrena dorsata, Andrena flavipes, Andrena haemorrrhoa, Andrena nitida; **Halictidae**: Halictus cochlearitarsis, Halictus confusus, Halictus patellatus, Halictus quadricinctus, Halictus senilis, Seladonia tumulorum, Lasioglossum calceatum
Beetles: **Scarabaeidae**: Cetonia aurata, Tropinota hirta, Tropinosa squalida Butterflies: **Papilionidae:** Iphiclides podalirius

Orchis simia Lam. 1779 subsp. *simia*
Pollinated by long-tongued bees (Apidae), moths (Sphingidae) and beetles (Elateridae, Glaphyridae, Scarabaeidae).
Long-tongued bees*:* **Apidae*:*** Apis mellifera, Eucera longicornis Moths: **Sphingidae:** Hemaris fuciformis
Beetles: **Elateridae**: Cidnopus pilosus; **Glaphyridae**: Amphicoma iberica; **Scarabaeidae**: Tropinota hirta

Orchis spitzelii subsp. *cazorlensis* (Lacaita) D. Riveera & López Vélez, 1987
Pollinated by long-tongued bees (Apidae).
Long-tongued bees: **Apidae:** Bombus barbutellus, Bombus lapidarius, Bombus pascuorum, Bombus pratorum, Bombus terrestris

Orchis spitzelii Saut. ex W.D.J. Koch 1837 subsp. *spitzelii*
Pollinated by long-tongued bees (Apidae).
Long-tongued bees: **Apidae:** Bombus bohemicus, Bombus campestris, Bombus lapidarius, Bomus lucorum, Bombus pascuorum, Bombus rupestris,

Orchis tenera (Landwehr) Kreutz 1991
Pollinated by long-tongued bees (Apidae, Megachilidae).
Long-tongued bees: **Apidae:** Apis mellifera, Bombus hortorum, Bombus lapidarius, Bombus lucorum, Bombus pascuorum, Bombus pratorum, Bombus ruderarius, Bombus ruderatus, Bombus rupestris, Bombus subterraneus, Bombus sylvarum, Bombus sylvestris, Bombus terrestris, Eucera hungarica, Eucera interrupta, Eucera longicornis; **Megachilidae**: Osmia bicolor

Platanthera algeriensis Batt. & Trab. 1892
Pollinated by butterflies (Lycaenidae, Papilionidae) and moths (Geometridae, Noctuidae, Sphingidae), long-tongued bees (Apidae) and flies (Calliphoridae).
Butterflies: **Lycaenidae**: Satyrium w-album; **Papilionidae:** Papilion machaon
Moths: **Geometridae:** Aplocera plagiata, Gnophos obfuscata, Scopula ornata, Xanthorrhoea ferrugata; **Noctuidae**: Agrotis puta, Apamea furva, Autographa gamma, Hada plebeja, Noctua pronuba, Orthosia incerta; **Sphingidae:** Deilephila elpenor, Deilephila porcellus
Long-tongued bees: **Apidae:** Apis mellifera, Bombus lapidarius Flies: **Calliphoridae**: Calliphora vicina

Platanthera bifolia (L.) Rich. 1817 subsp. *bifolia*
Pollinated by butterflies (Papilionidae) and moths (Geometridae, Noctuidae, Sphingidae), long-tongued bees (Apidae) and flies (Culicidae).
Butterflies: **Papilionidae:** Papilion machaon
Moths; **Geometridae:** Aplocera plagiata, Campaea margaritaria, Dysstroma citrata, Gnophos furvata, Gnophos obfuscata, Hydriomena furcata, Mesoleuca albicillata, Scopula virgulata, Xanthorhoea fluctuata; **Noctuidae:** Agrotis ipsilon, Agrotis segetum, Apamea furva, Apamea monoglypha, Autographa gamma, Autographa pulchrina, Cucullia umbrática, Hada plebeja, Hadena albimacula, Hadena expectata, Hadena silenes, Noctua pronuba, Orthosia incerta, Plusia festucae, Sideridis reticulata; **Sphingidae:** Deilephila elpenor, Deilephila porcellus, Hemaris fuciformis, Hyles euphorbiae, Sphinx pinastri.
Long-tongued bees: **Apidae:** Apis mellifera, Bombus terrestris Flies: **Culicidae:** Aedes communis, Aedes vexans

Platanthera clorantha Cust. ex Rchb. 1829
Pollinated by butterflies (Hesperiidae, Papilionidae), moths (Drepanidae, Geometridae, Noctuidae, Pyralidae, Sphingidae) and long-tongued bees (Apidae).
Butterflies: **Hesperiidae:** Ochlodes sylvanus; **Papilionidae**: Papilion machaon
Moths: **Drepanidae:** Habrosyne pyritoides; **Geometridae:** Aplocera plagiata, Gnophos obfuscata, Ourapteryx sambucaria, Scopula imitaria; **Noctuidae**: Abrostola tripartita, Abrostola triplasia, Agrotis cinerea, Apamea anceps, Apamea furva, Apamea lateritia, Apamea monoglypa, Apamea sublustris, Autographa bractea, Autographa gamma, Autographa jota, Autographa pulchrina, Cucullia umbrática, Diachrysia chrysitis, Diarsia mendica, Dicharyris forcipula, Hadena albimacula, Hadena caesia, Hadena compta, Hadena perplexa, Heliothis nubigera, Macdunnoughia confusa, Noctua prónuba, Plusia festucae, Polia bombycina, Polia hepatica, Polia nebulosa, Pyrrhia umbra, Rivula sericealis, Sideridis reticulata; **Pyralidae**: Pyralis regalis; **Sphingidae**: Agrius convolvuli, Deilephila elpenor, Deilephila porcellus, Hyles gallii, Proserpinus proserpina, Sphinx ligustri, Sphinx pinastri.
Long-tongued bees: **Apidae:** Apis mellifera, Bombus hortorum

Pseudorchis albida (l.) A. Löve & D. Löve

Pollinated by butterflies (Nymphalidae), moths (Crambidae, Plutellidae, Pterophoridae) and flies (Empididae).
Butterflies: **Nymphalidae:** Coenonympha pamphillus
Moths: **Crambidae**: Crambus pascuella, Crambus perlella, Chrysoteuchia culmella, Udea uliginosallis; **Plutellidae**: Plutella xylostella; **Pterophoridae**: Hellinsia osteodactylus, Hellinsia pectodactylus
Flies: **Empididae:** Empis bistortae, Empis tessellata

Serapias cordigera L. 1763 subsp. *cordigera*
Pollinated by long-tongued bees (Apidae, Megachilidae).
Long-tongued bees: **Apidae**: Anthophora mucida, Eucera clypeata, Eucera collaris; **Megachilidae:** Anthidium manicatum, Dioxys cincta, Hoplitis acuticornis ssp. hispanica, Hoplitis adunca, Hoplitis annulata, Hoplitis anthocopoides, Hoplitis benoisti, Hoplitis cristatula, Megachile albisecta, Megachile argentata, Megachile centuncularis, Megachile ericetorum, Megachile octosignata, Megachile pyrenaica, Megachile versicolor, Megachile willughbiella, Osmia argyropyga, Osmia aurulenta, Osmia bicolor, Osmia brevicornis, Osmia caerulescens, Osmia cephalotes, Osmia niveata, Osmia rufohirta, Osmia scutellaris, Osmia signata, Osmia submicans

Serapias lingua subsp. *duriaei* (Rchb. ex Batt.) Soó 1928 **Pollinated by long-tongued bees (Apidae)** Long-tongued bees: **Apidae**: Ceratina cucurbitina

Serapias lingua L. 1753 subsp. *lingua*
Pollinated by long-tongued bees (Apidae) and wasps (Crabronidae) Long-tongued bees: **Apidae**: Ceratina cucurbitina, Xylocopa iris Wasps: **Crabronidae**: Gorytes quinquecinctus

Serapias maria F.M. Vázquez 2008
Pollinated by long-tongued bees (Apidae).
Long-tongued bees: **Apidae**: Eucera longicornis

Serapias nurrica Corrias 1982
Pollinated by long-tongued bees (Megachilidae).
Long-tongued bees**: Megachilidae:** Osmia bicornis

Serapias occidentalis C. Venhuis & P. Venhuis 2006 **Pollinated by Long-tongued bees (Apidae)** Long-tongued bees**: Apidae:** Eucera longicornis

Serapias olbia Verg. 1908
Pollinated by long-tongued bees (Apidae).
Long-tongued bees: **Apidae**: Ceratina cucurbitina

Serapias parviflora Parl. 1837
Pollinated by long-tongued bees (Apidae, Megachilidae).
Long-tongued bees**: Apidae:** Ceratina cucurbitina; **Megachilidae:** Anthidium manicatum, Megachile willughbiella

Serapias perez-chiscanoi Acedo 1990
Pollinated by long-tongued bees (Apidae).
Long-tongued bees**: Apidae:** Bombus terrestris

Serapias strictiflora Wellw. ex Veiga 1887 **Pollinated by long-tongued bees (Apidae)**
Long-tongued bees**: Apidae:** Eucera collaris

Serapias vomeracea Briq. 1910
Pollinated by long-tongued bees (Apidae, Megachilidae), short-tongued bees (Halictidae, Andrenidae), beetles (Lymexylidae, Oedemeriae, Scarabaeidae) and wasps (Chrysididae).
Long-tongued bees: **Apidae**: Ceratina chalybea, Eucera clypeata, Eucera collaris, Eucera longicornis, Eucera seminuda; **Megachilidae**: Anthidium cingulatum, Anthidium manicatum, Hoplitis adunca, Hoplitis lepeletieri, Megachile centuncularis, Megachile melanopyga, Megachille willughbiella, Osmia aurulenta, Osmia leaiana, Osmia niveata, Rhodanthidium septemdentatum Short-tongued bees: **Halictidae:** Halictus scabiosae; **Andrenidae:** Andrena morio Beetles: **Lymexylidae:** Elateroides dermestoides**; Oedemeridae**: Oedemera nobilis; **Scarabaeidae**: Hoplia philanthus, Oxythera funesta.
Wasps: **Chrysididae**: Chrysura refulgens

Spiranthes aestivalis (Poir.) Rich. 1817
Pollinated by long-tongued bees (Apidae, Megachilidae), short-tongued bees (Halictidae, Colletidae), flies (Syrphidae), butterflies (Hesperiidae, Pieridae) and moths (Noctuidae).
Long-tongued bees: **Apidae**: Anthophora plumipes, Apis mellifera, Bombus pascuorum, Bombus pratorum, Bombus terrestris, Ceratina cucurbitina; **Megachilidae**: Anthidium manicatum, Megachille willughbiella, Osmia bicolor, Stelis punctulatissima
Short-tongued bees: **Halictidae**: Halictus simplex, Lasioglossum morio, Sphecodes gibbus; **Colletidae:** Hylaeus communis
Flies: **Syrphidae**: Sphaerophoria scripta
Butterflies: **Hesperiidae**: Thymelicus lineola; **Nymphalidae**: Vanessa carduia; **Pieridae**: Pieris rapae
Moths: **Noctuidae:** Syngrapha interrogationis

Spiranthes spiralis (L.) Chevall 1827
Pollinated by long-tongued bees (Apidae, Megachilidae) and short-tongued bees (Halictidae, Andrenidae), butterflies (Hesperiidae, Nymphalidae, Pieridae), moths (Noctuidae) and flies (Syrphidae).
Long-tongued bees: **Apidae**: Apis mellifera, Bombus lapidarius, Bombus pascuorum, Bombus sylvarum, Bombus terrestris; **Megachilidae**: Hoplitis adunca, Megachile parietina, Osmia bicornis Short-tongued bees: **Halictidae**: Lasioglossum zonulum, Halictus simplex; **Andrenidae**: Andrena nigroaenea
Butterflies: **Hesperiidae**: Erynnis tages; **Nymphalidae**: Vanessa atalanta; **Pieridae**: Pieris brassicae Moths: **Noctuidae:** Syngrapha interrogationis
Flies: **Syrphidae:** Episyrphus balbeatus

HYBRIDS

INTRAGENIC NOTOTAXONS

— Epipactis ×cardonneae J.M. Lewin (Epipactis atrorubens × Epipactis kleinii)
Pollinated by wasps (Vespidae)
Wasps: **Vespidae**: Dolichovespula saxonica, Dolichovespula sylvestris

— Epipactis ×vinturensis R. Martin (Epipactis atrorubens × Epipactis tremolsii)
Pollinated by wasps (Vespidae, Crabronidae)
Wasps: **Vespidae**: Dolichovespula norvegica, Dolichovespula saxonica, Dolichovespula sylvestris, Polistes biglumis, Vespula austriaca, Vespula germánica, Vespula rufa, Vespula vulgaris;

— Epipactis cardina × Epipactis distans **Pollinated by wasps (Vespidae, Crabronidae)**
Wasps: **Vespidae**: Dolichovespula sylvestris

— Epipactis cardina × Epipactis helleborine
Pollinated by wasps (Vespidae, Crabronidae)
Wasps: **Vespidae**: Dolichovespula media, Dolichovespula saxonica, Dolichovespula sylvestris, Polistes bischofii, Vespa crabro, Vespula austriaca, Vespula germánica, Vespula rufa, Vespula vulgaris

— Epipactis ×conquensis B. Ayuso & C. E. Hermos. Ayuso & C.E. Hermos (Epipactis cardina × Epipactis kleinii)
Pollinated by wasps (Vespidae, Crabronidae)
Wasps: **Vespidae**: Dolichovespula media, Dolichovespula saxonica, Dolichovespula sylvestris, Polistes bischofii, Polistes dominula, Vespa crabro, Vespula austriaca, Vespula rufa, Vespula vulgaris

— Epipactis cardina × Epipactis provincialis
Pollinated by wasps (Vespidae)
Wasps: **Vespidae:** Dolichovespula saxonica, Vespula vulgaris

— Epipactis distans × Epipactis helleborine
Pollinated by wasps (Vespidae)
Wasps: **Vespidae**: Dolichovespula media, Dolichovespula norvegica, Dolichovespula saxonica, Dolichovespula sylvestris, Polistes dominula, Vespa crabro, Vespula austriaca, Vespula germanica, Vespula rufa, Vespula vulgaris

— Epipactis distans × Epipactis kleinii
Pollinated by wasps (Vespidae)
Wasps: **Vespidae**: Dolichovespula media, Dolichovespula saxonica

— Epipactis fageticola × Epipactis helleborine
Pollinated by wasps (Vespidae)
Wasps: **Vespidae**: Dolichovespula media, Dolichovespula norvegica, Dolichovespula saxonica, Dolichovespula sylvestris, Polistes dominula, Vespa crabro, Vespula austriaca, Vespula germanica, Vespula vulgaris

— Epipactis ×reinekei M. Bayer (Epipactis helleborine × Epipactis muelleri)

Pollinated by wasps (Vespidae)
Wasps: **Vespidae**: Dolichovespula norvegica, Dolichovespula saxonica, Dolichovespula sylvestris, Polistes dominula, Vespa crabro, Vespula germánica, Vespula rufa, Vespula vulgaris

— Epipactis helleborine × Epipactis phyllanthes
Pollinated by wasps (Vespidae)
Wasps: **Vespidae**: Dolichovespula media, Dolichovespula norvegica, Dolichovespula saxonica, Dolichovespula sylvestris, Polistes bischofii, Polistes dominula, Vespula austriaca, Vespula germánica, Vespula rufa, Vespula vulgaris

— Epipactis ×gevaudanii P. Delforge (Epipactis helleborine × Epipactis rhodanensis)
Pollinated by wasps (Vespidae)
Wasps: **Vespidae**: Dolichovespula media, Dolichovespula norvegica, Dolichovespula saxonica, Dolichovespula sylvestris, Polistes dominula, Vespula austriaca, Vespa crabro, Vespula germánica, Vespula rufa, Vespula vulgaris

- Epipactis ×populetorum B. Ayuso & C.E. Hermos (Epipactis campeadorii × Epipactis helleborine)
Pollinated by wasps (Vespidae)
Wasps: **Vespidae**: Dolichovespula media, Dolichovespula norvegica, Dolichovespula saxonica, Dolichovespula sylvestris, Polistes dominula, Vespula austriaca, Vespa crabro, Vespula rufa, Vespula vulgaris

— Epipactis kleinii × Epipactis tremolsii
Pollinated by wasps (Vespidae).
Wasps: **Vespidae**: Dolichovespula norvegica, Dolichovespula saxonica, Dolichovespula sylvestris, Polistes biglumis, Polistes dominula, Vespula austriaca, Vespula germánica, Vespula rufa, Vespula vulgaris;

— Epipactis lusitanica × Epipactis tremolsii
Pollinated by wasps (Vespidae, Crabronidae).
Wasps: **Vespidae**: Dolichovespula norvegica, Dolichovespula saxonica, Dolichovespula sylvestris, Polistes biglumis, Polistes dominula, Vespula austriaca, Vespula germánica, Vespula rufa, Vespula vulgaris;

— Epipactis microphylla × Epipactis tremolsii
Pollinated by wasps (Vespidae).
Wasps: **Vespidae**: Dolichovespula norvegica, Dolichovespula saxonica, Dolichovespula sylvestris, Eumenes papillarius, Euodynerus fastidiosus, Polistes biglumis, Polistes dominula, Vespula austriaca, Vespula rufa, Vespula vulgaris;

— Epipactis phyllanthes × Epipactis rhodanensis
Pollinated by wasps (Vespidae).
Wasps: **Vespidae**: Vespula vulgaris.

— Epipactis rhodanensis × Epipactis kleinii
Pollinated by wasps (Vespidae).
Wasps: **Vespidae**: Dolichovespula sylvestris

- Epipactis campeadorii × Epipactis parviflora

Pollinated by wasps (Vespidae).

Wasps: **Vespidae**: Vespula rufa

— Platanthera ×hybrida Brügger (Platanthera bifolia × Platanthera chlorantha)

Pollinated by moths (Noctuidae, Sphingidae)

Noctuidae: Abrostola asclepiadis, Abrostola tripartita, Abrostola triplasia, Agrotis bigramma, Agrotis cinerea, Agrotis fatidica, Agrotis ipsilon, Agrotis lata, Agrotis obesa, Agrotis puta, Apamea anceps, Apamea furva, Apamea lateritia, Apamea monoglypa, Apamea sublustris, Autographa bractea, Autographa gamma, Autographa jota, Autographa pulchrina, Cucullia lucifuga, Cucullia umbratica, Diachrysia chrysitis, Diarsia mendica, Dicharyris forcipula, Hadena albimacula, Hadena caesia, Hadena compta, Hadena confusa, Hadena luteocincta, Hadena perplexa, Hadena sancta, Heliothis nubigera, Macdunnoughia confusa, Noctua comes, Noctua fimbriata, Noctua interjecta.

Noctua janthe, Noctua janthina, Noctua pronuba, Plusia festucae, Polia bombycina, Polia hepatica, Polia bombycina, Polia hepatica, Polia nebulosa, Polia sagittigera, Pyrrhia umbra, Rivula sericealis, Sideridis reticulata; **Sphingidae:** Deilephila elpenor, Deilephila porcellus, Hemaris fuciformis, Hyles euphorbiae, Sphinx pinastri.

— Gymnadenia conopsea × Gymnadenia densiflora

Pollinated by butterflies (Hesperiidae, Lycaenidae, Nymphalidae, Papilionidae, Pieridae) and moths (Crambidae, Geometridae, Micropterigidae, Noctuidae, Pterophoridae Pyralidae, Sphingidae, Zygaenidae).

Butterflies: **Hesperiidae**: Ochlodes sylvanus, Thymelicus lineola, Thymelicus sylvestris; **Lycaenidae:** Lycaena phlaeas, Plebejus idas, Polyommatus icarus; **Nymphalidae**: Aglais urticae, Argynis adipe, Argynnis aglaja, Boloria euphrosyne, Euphrydryas aurinia, Manola jurtina, Vanessa cardui; **Papilionidae**: Papilio machaon; **Pieridae**: Aporia crategi, Pieris rapae.

Moths: **Crambridae:** Crambus perlella; **Geometridae:** Aplocera plagiata, Entephria caesiata, Heliothea discoidaria, Siona lineata; **Noctuidae**: Autographa aemula, Autographa bractea, Autographa gamma, Autographa jota, Autographa pulchrina, Mythimna conigera, Mythimna prominens, Mythimna sicula, Mythimna turca, Mythimna unipuncta, Mythimna vitellina, Papestra biren, Phytometra viridaria; **Sphingidae**: Deilephela porcellus, Hemaris tytius, Hyles euphorbiae, Hyles galii, Hyles livornica, Macroglossum stellatarum, Sphinx ligustri, Sphinx pinastri; **Zygaenidae**: Adscita geryon, Adscita statites, Zygaena exulans, Zygaena filipendulae, Zygaena lonicerae, Zygaena loti, Zygaena trifolii.

— Gymnadenia conopsea × Gymnadenia gabasiana

Pollinated by butterflies (Hesperiidae, Lycaenidae, Nymphalidae, Papilionidae, Pieridae) and moths (Crambridae, Geometridae, Noctuidae, Sphingidae, Zygaenidae).

Butterflies: **Hesperiidae**: Ochlodes sylvanus, Tymelicus sylvestris; **Lycaenidae:** Plebejus idas, Polyommatus icarus; **Nymphalidae**: Aglais urticae, Vanessa cardui; **Papilionidae**: Papilio machaon; **Pieridae**: Aporia crategi.

Moths: **Crambridae:** Pyrausta despicata; **Geometridae:** Aplocera plagiata, Entephria caesiata, Myinodes interpunctaria, Siona lineata; **Noctuidae**: Autographa bractea, Autographa gamma, Autographa jota, Autographa pulchrina, Mythimna congrua, Mythimna conigera, Mythimna l- album, Mythimna langida, Mythimna pudorina, Papestra biren, Phytometra viridaria; **Sphingidae**: Deilephela porcellus, Hemaris tytius, Hyles euphorbiae, Hyles galii,

Hyles livornica, Macroglossum stellatarum, Sphinx ligustri, Sphinx pinastri; **Zygaenidae**: Adscita geryon, Adscita statites, Zygaena exulans, Zygaena filipendulae, Zygaena lonicerae, Zygaena loti, Zygaena trifolii.

- Gymnadenia ×pyrenaeensis Foelsche; ×Gymnigritella pyrenaica C.E. Hermosilla & Sabando **Pollinated by butterflies (Hesperiidae, Nymphalidae, Pieridae) and moths (Geometridae, Noctuidae, Sphingidae, Zygaenidae).**

Butterflies: **Hesperiidae**: Ochlodes sylvanus, Tymelicus sylvestris; **Lycaenidae:** Polyommatus icarus; **Nymphalidae**: Aglais urticae, Maniola jurtina, Vanessa cardui; **Papilionidae**: Papilio machaon; **Pieridae**: Aporia crategi, Pieris brassicae.
Moths: **Geometridae:** Entephria caesiata, Eupithecia alliaria, Eupithecia extensaria, Myinodes interpunctaria, Siona lineata; **Noctuidae**: Autographa bractea, Autographa gamma, Autographa jota, Autographa jota, Autographa pulchrina, Mythimna congrua, Mythimna conigera, Mythimna l- album, Mythimna langida, Mythimna pudorina, Mythimna vitellina, Papestra biren, Phytometra viridaria; **Sphingidae**: Deilephela porcellus, Hemaris tytius, Hyles euphorbiae, Hyles galii, Hyles livornica, Macroglossum stellatarum, Sphinx ligustri, Sphinx pinastri; **Zygaenidae**: Adscita geryon, Adscita statites, Zygaena exulans, Zygaena filipendulae, Zygaena lonicerae, Zygaena loti, Zygaena purpuralis, Zygaena sarpedon, Zygaena trifolii.

— Gymnadenia ×intermedia Peterm. (Gymnadenia conopsea × Gymnadenia odoratissima)
Pollinated by butterflies (Hesperiidae, Nymphalidae, Pieridae) and moths (Geometridae, Noctuidae, Sphingidae, Zygaenidae).
Butterflies: **Hesperiidae**: Ochlodes sylvanus, Tymelicus sylvestris; **Lycaenidae:** Polyommatus icarus; **Nymphalidae**: Aglais urticae, Melithaea athalia, Vanessa cardui; **Pieridae**: Aporia crategi, Pieris brassicae.
Moths: **Geometridae:** Entephria caesiata, Eupithecia centaurata, Eupithecia dodoneata, Idaea aversata, Myinodes interpunctaria, Siona lineata; **Noctuidae**: Autographa bractea, Autographa gamma, Autographa jota, Autographa jota, Autographa pulchrina, Mythimna congrua, Mythimna conigera, Mythimna l-album, Mythimna langida, Mythimna pudorina, Mythimna vitellina, Noctua prónuba, Papestra biren, Phytometra viridaria; **Sphingidae**: Deilephela porcellus, Hemaris tytius, Hyles euphorbiae, Hyles galii, Hyles livornica, Macroglossum stellatarum, Sphinx ligustri, Sphinx pinastri; **Zygaenidae**: Adscita geryon, Adscita statites, Zygaena exulans, Zygaena filipendulae, Zygaena lonicerae, Zygaena loti, Zygaena purpuralis, Zygaena sarpedon, Zygaena trifolii.

— Gymnadenia conopsea × Gymnadenia pyrenaica
Pollinated by butterflies (Hesperiidae, Nymphalidae, Pieridae) and moths (Geometridae, Noctuidae, Sphingidae, Zygaenidae).
Butterflies: **Hesperiidae**: Ochlodes sylvanus, Thymelicus lineola, Tymelicus sylvestris; **Lycaenidae:** Lycaena phlaeas, Polyommatus icarus; **Nymphalidae**: Aglais urticae, Coenonimpha pamphilus, Maniola jurtina, Vanessa cardui; **Pieridae**: Aporia crategi, Pieris napi, Pieris rapae.
Moths: **Geometridae:** Apamea furva, Eupithecia graphata, Idaea biselata, Idaea biselata, Myinodes interpunctaria, Siona lineata; **Noctuidae**: Autographa bractea, Autographa gamma, Autographa jota, Autographa jota, Autographa pulchrina, Mythimna l-album, Mythimna langida, Mythimna pudorina, Mythimna unipunctata, Mythimna vitellina, Papestra biren, Phytometra viridaria; **Sphingidae**: Deilephela porcellus, Hemaris tytius, Hyles euphorbiae,

Hyles galii, Hyles livornica, Macroglossum stellatarum, Sphinx ligustri, Sphinx pinastri; **Zygaenidae**: Adscita geryon, Adscita statites, Zygaena exulans, Zygaena filipendulae, Zygaena lonicerae, Zygaena loti, Zygaena trifolii, Zygaena viciae.

- Gymnadenia ×proxima C.E. Hermosa (Gymnadenia conopsea × Gymnadenia odoratissima subsp. longicalcarata)

Pollinated by butterflies (Hesperiidae, Nymphalidae, Pieridae) and moths (Geometridae, Noctuidae, Sphingidae, Zygaenidae).

Butterflies: **Hesperiidae**: Ochlodes sylvanus, Thymelicus lineola, Tymelicus sylvestris; **Lycaenidae:** Lycaena phlaeas, Polyommatus icarus; **Nymphalidae**: Aglais urticae, Coenonimpha pamphilus, Maniola jurtina, Vanessa cardui; **Pieridae**: Aporia crategi, Pieris napi, Pieris rapae.

Moths: **Geometridae:** Apamea furva, Eupithecia graphata, Idaea biselata, Idaea biselata, Myinodes interpunctaria, Siona lineata; **Noctuidae**: Autographa bractea, Autographa gamma, Autographa jota, Autographa jota, Autographa pulchrina, Mythimna l-album, Mythimna langida, Mythimna pudorina, Mythimna unipunctata, Mythimna vitellina, Papestra biren, Phytometra viridaria; **Sphingidae**: Deilephela porcellus, Hemaris tytius, Hyles euphorbiae, Hyles galii, Hyles livornica, Macroglossum stellatarum, Sphinx ligustri, Sphinx pinastri; **Zygaenidae**: Adscita geryon, Adscita statites, Zygaena exulans, Zygaena filipendulae, Zygaena lonicerae, Zygaena loti, Zygaena trifolii, Zygaena viciae.

— Gymnadenia densiflora × Gymnadenia pyrenaica

Pollinated by butterflies (Hesperiidae, Nymphalidae, Pieridae) and moths (Geometridae, Noctuidae, Sphingidae, Zygaenidae).

Butterflies: **Hesperiidae**: Ochlodes sylvanus, Thymelicus lineola, Tymelicus sylvestris; **Lycaenidae:** Lycaena phlaeas, Polyommatus icarus; **Nymphalidae**: Aglais urticae, Coenonimpha pamphilus, Maniola jurtina, Vanessa cardui; **Pieridae**: Aporia crategi, Pieris napi, Pieris rapae.

Moths: **Geometridae:** Apamea furva, Eupithecia graphata, Idaea biselata, Idaea biselata, Myinodes interpunctaria, Siona lineata; **Noctuidae**: Autographa bractea, Autographa gamma, Autographa jota, Autographa jota, Autographa pulchrina, Mythimna l-album, Mythimna langida, Mythimna pudorina, Mythimna unipunctata, Mythimna vitellina, Papestra biren, Phytometra viridaria; **Sphingidae**: Deilephela porcellus, Hemaris tytius, Hyles euphorbiae, Hyles galii, Hyles livornica, Macroglossum stellatarum, Sphinx ligustri, Sphinx pinastri; **Zygaenidae**: Adscita geryon, Adscita statites, Zygaena exulans, Zygaena filipendulae, Zygaena lonicerae, Zygaena loti, Zygaena trifolii, Zygaena viciae.

(Gymnadenia densiflora × Gymnadenia odoratissima subsp. longicalcarata) -Gymnadenia ×sabandoi C.E. Hermos (Gymnadenia densiflora × Gymnadenia odoratissima subsp. longicalcarata)

Pollinated by butterflies (Hesperiidae, Nymphalidae, Pieridae) and moths (Geometridae, Noctuidae, Sphingidae, Zygaenidae).

Butterflies: **Hesperiidae**: Ochlodes sylvanus, Thymelicus lineola, Tymelicus sylvestris; **Lycaenidae:** Lycaena phlaeas, Polyommatus icarus; **Nymphalidae**: Aglais urticae, Coenonimpha pamphilus, Maniola jurtina, Vanessa cardui; **Pieridae**: Aporia crategi, Pieris napi, Pieris rapae.

Moths: **Geometridae:** Apamea furva, Eupithecia graphata, Idaea biselata, Idaea biselata, Myinodes interpunctaria, Siona lineata; **Noctuidae**: Autographa bractea, Autographa gamma, Autographa jota, Autographa jota, Autographa pulchrina, Mythimna l-album, Mythimna langida, Mythimna pudorina, Mythimna unipunctata, Mythimna vitellina, Papestra biren, Phytometra viridaria; **Sphingidae**: Deilephela porcellus, Hemaris tytius, Hyles euphorbiae, Hyles galii, Hyles livornica, Macroglossum stellatarum, Sphinx ligustri, Sphinx pinastri; **Zygaenidae**: Adscita geryon, Adscita statites, Zygaena exulans, Zygaena filipendulae, Zygaena lonicerae, Zygaena loti, Zygaena trifolii, Zygaena viciae.

— Dactylorhiza alpestris × Dactylorhiza fuchsii; Dactylorhiza ×sooi (Ruppert ex Soó) Soó
Pollinated by bumblebees. Bombus lapidarius, Bombus pascuorum, Bombus terrestris.

— Dactylorhiza caramulensis × Dactylorhiza saccifera?
Pollinated by bumblebees. Bombus lapidarius, Bombus pascuorum, Bombus terrestris.

— Dactylorhiza elata × Dactylorhiza ericetorum ?
Pollinated by bumblebees. Bombus lapidarius, Bombus pascuorum, Bombus terrestris.

— Dactylorhiza ×hjerstonii P.P. Ferrer, Lozano, Roselló, Feliu & Peña (Dactylorhiza elata × Dactylorhiza fuchsii)
Pollinated by bumblebees. Bombus lapidarius, Bombus pascuorum, Bombus terrestris.

— Dactylorhiza ×dubreuilhii (G. Keller & Jeanj.) Soó (Dactylorhiza elata × Dactylorhiza incarnata) Pollinated by bumblebees. Bombus lapidarius, Bombus pascuorum, Bombus terrestris.

— Dactylorhiza ×perez-chiscanoi F.M. Vázquez (Dactylorhiza elata × Dactylorhiza irenica?) Pollinated by bumblebees. Bombus lapidarius, Bombus pascuorum, Bombus terrestris.

—; Dactylorhiza ×delamainii Keller & Stephenson (Dactylorhiza elata × Dactylorhiza maculata) Pollinated by bumblebees. Bombus lapidarius, Bombus pascuorum, Bombus terrestris.

— Dactylorhiza elata × Dactylorhiza majalis
Pollinated by bumblebees. Bombus lapidarius, Bombus pascuorum, Bombus terrestris.

— Dactylorhiza elodes? × Dactylorhiza purpurella?
Pollinated by bumblebees. Bombus lapidarius, Bombus pascuorum, Bombus terrestris.

— Dactylorhiza fuchsii × Dactylorhiza incarnata; Dactylorhiza ×wirtgenii Höppner; Dactylorhiza
×kerneriorum (Soó) Soó
Pollinated by bumblebees. Bombus lapidarius, Bombus pascuorum, Bombus terrestris.

— Dactylorhiza fuchsii × Dactylorhiza maculata; Dactylorhiza ×transiens (Druce) Soó (ut Dactylorhiza ericetorum × Dactylorhiza fuchsii)

Pollinated by bumblebees. Bombus lapidarius, Bombus pascuorum, Bombus terrestris.

— Dactylorhiza fuchsii × Dactylorhiza majalis; Dactylorhiza ×braunii (Halácsy) Borsos & Soó — Dactylorhiza guimaraesi × Dactylorhiza sambucina ¿? (ut Dactylorhiza markusii × Dactylorhiza sambucina).

Pollinated by bumblebees. Bombus lapidarius, Bombus pascuorum, Bombus terrestris.

— Dactylorhiza incarnata × Dactylorhiza maculata; Dactylorhiza ×carnea (E.G. Camús) Soó; Dactylorhiza ×maculatiformis Rouy; Dactylorhiza ×ambigua Kern.; Dactylorhiza claudiopolitana (Simonk.) Borsos & Soó.
Pollinated by bumblebees. Bombus lapidarius, Bombus pascuorum, Bombus terrestris.

— Dactylorhiza incarnata × Dactylorhiza majalis; Dactylorhiza ×aschersoniana (Hausskn.) Borsos.
& Soó; Dactylorhiza ×pseudotransteuneri Fuchs; Dactylorhiza ×gennachensis Fuchs; Dactylorhiza ×gennachensis Fuchs; Dactylorhiza ×koningweeniana Fuchs; Dactylorhiza ×eifliaca Fuchs; Dactylorhiza ×pseudojunialis R. Doll

Pollinated by bumblebees. Bombus lapidarius, Bombus pascuorum, Bombus terrestris.

— Dactylorhiza insularis × Dactylorhiza sambucina; Dactylorhiza ×cantabrica H.A. Pedersen
Pollinated by bumblebees. Bombus lapidarius, Bombus pascuorum, Bombus terrestris.

— Dactylorhiza maculata × Dactylorhiza majalis; Dactylorhiza ×dinglensis (Wilmott) Soó; Dactylorhiza ×vermeuleniana Soó
Pollinated by bumblebees. Bombus lapidarius, Bombus pascuorum, Bombus terrestris.

— Orchis canariensis × Orchis lapalmensis; Orchis ×reginae Leibbach & Ruedi Peter
Pollinated by bumblebees. Bombus canariensis.

— Orchis cazorlensis × Orchis langei; Orchis ×incantata P. Delforge
Pollinated by bumblebees. Bombus lapidarius, Bombus pascuorum, Bombus terrestris.

— Orchis cazorlensis × Orchis mascula
Pollinated by bumblebees. Bombus lapidarius, Bombus pascuorum, Bombus terrestris.

— Orchis champagneuxii × Orchis collina; Orchis ×semisaccata E.G. Camús; Orchis ×semichampagneuxii E.G. Camús; Orchis ×rainei Rouy pro sp.
Pollinated by bumblebees. Bombus lapidarius, Bombus pascuorum, Bombus terrestris.

— Orchis champagneuxii × Orchis coriophora
Pollinated by honey bees and bumblebees. Nectariferous plant. Apis mellifera, Bombus lapidarius, Bombus pascuorum, Bombus terrestris.

— Orchis champagneuxii × Orchis fragrans

Pollinated by honey bees and bumblebees. Nectariferous plant. Apis mellifera, Bombus lapidarius, Bombus pascuorum, Bombus terrestris.

– Orchis champagneuxii × Orchis laxiflora; Anacamptis ×rayyana Robles, Quintana & Becerra
Pollinated by bumblebees. Bombus lapidarius, Bombus pascuorum, Bombus terrestris.
– Orchis champagneuxii × Orchis morio; Orchis ×romerae Hervás

Pollinated by bumblebees. Bombus lapidarius, Bombus pascuorum, Bombus terrestris.

– Orchis champagneuxii × Orchis olbiensis; Orchis ×mezquitensis Pallarés
Pollinated by bumblebees. Bombus lapidarius, Bombus pascuorum, Bombus terrestris.

– Orchis champagneuxii × Orchis papilionacea; Orchis ×subpapilionacea R. Lopes
Pollinated by bumblebees. Bombus lapidarius, Bombus pascuorum, Bombus terrestris.

– Orchis champagneuxii × Orchis picta; Orchis ×albertii A. Camús
Pollinated by bumblebees. Bombus lapidarius, Bombus pascuorum, Bombus terrestris.

– Orchis collina × Orchis morio; Orchis ×murgiana (Medagli, D'Emerico, Ruggiero & Bianco) De Bellard, Hervás, Huertas & Reyes
Pollinated by bumblebees. Bombus lapidarius, Bombus pascuorum, Bombus terrestris.

– Orchis collina × Orchis olbiensis
Pollinated by bumblebees. Bombus lapidarius, Bombus pascuorum, Bombus terrestris.

– Orchis collina × Orchis papilionacea; Orchis ×dulukae Hautz; Orchis ×dafnii R. Luz & W. Schdmit

Pollinated by bumblebees. Bombus lapidarius, Bombus pascuorum, Bombus terrestris.

– Orchis conica × Orchis italica; Orchis ×diversifolia Gaudagno
Pollinated by bumblebees. Bombus lapidarius, Bombus pascuorum, Bombus terrestris.

– Orchis coriophora × Orchis morio; Orchis ×olida Bréb.; Orchis ×cimicina Bréb.
Pollinated by honey bees and bumblebees. Nectariferous plant. Apis mellifera, Bombus lapidarius, Bombus pascuorum, Bombus terrestris.

— Orchis coriophora × Orchis purpurea; Orchis ×celtiberica Pau
Pollinated by honey bees and bumblebees. Nectariferous plant. Apis mellifera, Bombus lapidarius, Bombus pascuorum, Bombus terrestris.

— Orchis fragrans × Orchis laxiflora; Orchis ×bicknellii E.G. Camús, Bergon & A. Camús; Orchis
×parvifolia Chaub.
Pollinated by honey bees and bumblebees. Nectariferous plant. Apis mellifera, Bombus lapidarius, Bombus pascuorum, Bombus terrestris.

— Orchis fragrans × Orchis papilionacea; Orchis ×menosii C. Bernard & G. Fabre

Pollinated by honey bees and bumblebees. Nectariferous plant. Apis mellifera, Bombus lapidarius, Bombus pascuorum, Bombus terrestris.

— Orchis fragrans × Orchis picta; Orchis ×pauliana Malinvaud; Orchis ×darcisii Murr
Pollinated by honey bees and bumblebees. Nectariferous plant. Apis mellifera, Bombus lapidarius, Bombus pascuorum, Bombus terrestris.

— Orchis fragrans × Orchis robusta; Orchis ×albuferensis R.M. Bateman & Hollingsworth
Pollinated by honey bees and bumblebees. Nectariferous plant. Apis mellifera,, Bombus lapidarius, Bombus pascuorum, Bombus terrestris.

— Orchis italica × Orchis purpurea; Orchis ×caesii De Angelis & Fumante

Pollinated by bumblebees. Bombus lapidarius, Bombus pascuorum, Bombus terrestris.

— Orchis langei × Orchis morio
Pollinated by bumblebees. Bombus lapidarius, Bombus pascuorum, Bombus terrestris.

— Orchis langei × Orchis olbiensis; Orchis ×serraniana P. Delforge
Pollinated by bumblebees. Bombus lapidarius, Bombus pascuorum, Bombus terrestris.

— Orchis langei × Orchis provincialis; Orchis ×navarrensis Amardeilh
Pollinated by bumblebees. Bombus lapidarius, Bombus pascuorum, Bombus terrestris.

— Orchis langei × Orchis tenera
Pollinated by bumblebees. Bombus lapidarius, Bombus pascuorum, Bombus terrestris.

— Orchis laxiflora × Orchis mascula; Orchis ×hispanica A. Nieschalk & C. Nieschalk
Pollinated by bumblebees. Bombus lapidarius, Bombus pascuorum, Bombus terrestris.

— Orchis laxiflora × Orchis morio; Orchis ×alata Fleury
Pollinated by bumblebees. Bombus lapidarius, Bombus pascuorum, Bombus terrestris.

— Orchis laxiflora × Orchis picta
Pollinated by bumblebees. Bombus lapidarius, Bombus pascuorum, Bombus terrestris.

— Orchis mascula × Orchis morio; Orchis ×vilmsii Camús; Orchis ×morioides Brand
Pollinated by bumblebees. Bombus lapidarius, Bombus pascuorum, Bombus terrestris.

— Orchis mascula × Orchis pallens; Orchis ×loreziana Bruegg.; Orchis ×haussknechtii M. Schulze
Pollinated by bumblebees. Bombus lapidarius, Bombus pascuorum, Bombus terrestris.

— Orchis mascula × Orchis provincialis; Orchis ×peuzegiana A. Camús & E.G. Camús
Pollinated by bumblebees. Bombus lapidarius, Bombus pascuorum, Bombus terrestris.

— Orchis mascula × Orchis tenera
Pollinated by bumblebees. Bombus lapidarius, Bombus pascuorum, Bombus terrestris.

— Orchis militaris × Orchis purpurea; Orchis ×hybrida Boenningh ex Rchb. fil.; Orchis ×jacquinii
Godr.; Orchis ×dubia E.G. Camús

Pollinated by bumblebees. Bombus lapidarius, Bombus pascuorum, Bombus terrestris.

— Orchis militaris × Orchis simia; Orchis ×beyrichii Kern; Orchis ×grenieri E.G. Camús; Orchis ×grenieri E.G. Camús; Orchis
×chatinii E.G. Camús; Orchis ×decipiens E.G. Camús; Orchis ×propinqua E.G. Camús & Bergon Pollinated by bumblebees. Bombus lapidarius, Bombus pascuorum, Bombus terrestris.

— Orchis morio × Orchis papilionacea; Orchis ×gennarii Rchb. fil.; Orchis ×debeauxii E.G. Camús ;
Orchis ×decipiens Bianca
Pollinated by bumblebees. Bombus lapidarius, Bombus pascuorum, Bombus terrestris.

— Orchis morio × Orchis picta; Orchis ×heraclea Verg.
Pollinated by bumblebees. Bombus lapidarius, Bombus pascuorum, Bombus terrestris.

— Orchis olbiensis × Orchis tenera

Pollinated by bumblebees. Bombus lapidarius, Bombus pascuorum, Bombus terrestris.

— Orchis papilionacea × Orchis picta; Orchis ×pseudopicta E.G. Camús, Bergon & A. Camús; Orchis ×pseudorubra Freyn ; Orchis ×yvesii Verg.; Orchis ×orientecaucasica B. Baumann, H. Baumann, R. Lorenz & R. Peter
Pollinated by bumblebees. Bombus lapidarius, Bombus pascuorum, Bombus terrestris.
— Orchis provincialis × Orchis tenera
Pollinated by bumblebees. Bombus lapidarius, Bombus pascuorum, Bombus terrestris.

— Orchis purpurea × Orchis simia; Orchis ×angusticruris Franch. ex Rouy; Orchis ×franchetii E.G. Camús; Orchis ×weddellii Franch; Orchis ×digenea Tourlet; Orchis ×gelmiana Dalla Torre & Sarnth. Pollinated by bumblebees. Bombus lapidarius, Bombus pascuorum, Bombus terrestris.

— Serapias cordigera × Serapias lingua; Serapias ×ambigua Rouy
Pollinated by Eucera clypeata, Eucera collaris;

— Serapias cordigera × Serapias maria; Serapias ×occidentalis C. Venhuis & P. Venhuis
Pollinated by Eucera clypeata, Eucera collaris;

— Serapias cordigera × Serapias parviflora; Serapias ×rainei E.G. Camús, Bergon & A. Camús;
Serapias ×alfredii Briq.
Pollinated by Eucera clypeata, Eucera collaris;

— Serapias cordigera × Serapias perez-chiscanoi Pollinated by Eucera clypeata, Eucera collaris;

— Serapias cordigera × Serapias strictiflora
Pollinated by Eucera clypeata, Eucera collaris;

— Serapias cordigera × Serapias vomeracea; Serapias ×kelleri A. Camús
Pollinated by Eucera clypeata, Eucera collaris;

— Serapias lingua × Serapias maria; Serapias ×liana F.M. Vázquez, A. Sánchez & G. Alonso

Pollinated by Eucera clypeata, Eucera collaris;

— Serapias lingua × Serapias parviflora; Serapias ×todaroi Tineo; Serapias ×semilingua E.G. Camús,

Bergon & A. Camús
Pollinated by Eucera clypeata, Eucera collaris;

— Serapias lingua × Serapias perez-chiscanoi; Serapias ×venhuisia F.M. Vázquez Pollinated by Eucera clypeata, Eucera collaris;

— Serapias lingua × Serapias strictiflora; Serapias ×stenopetala Maire & T. Stephenson Pollinated by Eucera clypeata, Eucera collaris;
— Serapias lingua × Serapias vomeracea; Serapias ×intermedia Forestier ex Rchb. fil. Pollinated by Eucera clypeata, Eucera collaris;

— Serapias parviflora × Serapias strictiflora
Pollinated by Eucera clypeata, Eucera collaris;

— Ophrys algarvensis × Ophrys atlantica Pollinated by Colletes albomaculatus

— Ophrys algarvensis × Ophrys fusca
Pollinated by Colletes albomaculatus, Andrena flavipes, Andrena nigroaenea, Andrena similis

— Ophrys algarvensis × Ophrys lupercalis; Ophrys ×pozoi E. Robles & M. Becerra
Pollinated by Colletes albomaculatus, Andrena nigroaenea

— Ophrys alpujata × Ophrys lupercalis
Pollinated by Andrena nigroaenea

— Ophrys apifera × Ophrys baleárica
Pollinated by Eucera longicornis, Megachile sicula

— Ophrys apifera × Ophrys bombyliflora; Ophrys ×circaea W. Rossi & G. Prola
Pollinated by Eucera longicornis, Eucera nigrescens

— Ophrys apifera × Ophrys catalaunica
Pollinated by Eucera nigrescens

— Ophrys apifera × Ophrys ficalhoana; Ophrys ×turiana J.E. Arnold
Pollinated by Eucera longicornis

— Ophrys apifera × Ophrys picta nafarroana; Ophrys ×pompelonensis E. Robles, M. Becerra & A.
Bece Pollinized by Eucera longicornis rra

— Ophrys apifera × Ophrys incubacea; Ophrys ×sandrae Sardu
Pollinated by Eucera longicornis

— Ophrys apifera × Ophrys scolopax; Ophrys ×pseudoscolopax Moggridge; Ophrys ×minuticauda
Duffort; Ophrys ×ouritensis Guitonneau Pollinated by Eucera longicornis

— Ophrys apifera × Ophrys speculum; Ophrys ×soller M. Henkel
Pollinated by Eucera longicornis

— Ophrys apifera × Ophrys tenthredinifera

Pollinated by Eucera longicornis

— Ophrys arachnitiformis × Ophrys bertolonii? Ophrys ×neocamusii Godfery
Pollinated by Andrena nigroaenea

— Ophrys arachnitiformis × Ophrys lupercalis; Ophrys ×carqueirannensis E.G. Camús

Pollinated by Andrena nigroaenea

— Ophrys arachnitiformis × Ophrys passionis
Pollinated by Andrena nigroaenea

— Ophrys arachnitiformis × Ophrys scolopax; Ophrys ×cranbrokeana Godfery

Pollinated by Andrena nigroaenea

— Ophrys arachnitiformis × Ophrys tenthredinifera; Ophrys ×laconensis Scrugli & M.P. Grasso
Pollinated by Andrena nigroaenea

— Ophrys araneola × Ophrys catalaunica
Pollinated by Andrena combinata

— Ophrys araneola × Ophrys sphegodes; Ophrys ×jeanpertii E.G. Camús
Pollinated by Andrena combinata

— Ophrys arnoldii × Ophrys fabrella ? Ophrys arnoldii × Ophrys fusca

Pollinated by Andrena nigroaenea

— Ophrys arnoldii × Ophrys lupercalis ¿? (ut Ophrys arnoldii × Ophrys forestieri)
Pollinated by Andrena nigroaenea

— Ophrys arnoldii × Ophrys passionis
Pollinated by Andrena nigroaenea

— Ophrys arnoldii × Ophrys speculum
Pollinated by Andrena nigroaenea

— Ophrys arnoldii × Ophrys sphegodes; Ophrys ×vistabellae J.E. Arnold
Pollinated by Andrena nigroaenea

— Ophrys arnoldii × Ophrys vasconica
Pollinated by Andrena nigroaenea

— Ophrys atlantica × Ophrys bilunulata
Pollinated by Megachile parietina

— Ophrys atlantica × Ophrys dyris; Ophrys ×kurzeorum H. Baumann
Pollinated by Megachile parietina

— Ophrys atlantica × Ophrys lupercalis; Ophrys ×joannae Maire; Ophrys ×angelicae Conesa nom.
Nudum
Pollinated by Megachile parietina

— Ophrys atlantica × Ophrys lutea; Ophrys ×migoutiana H. Gay

Pollinated by Megachile parietina

— Ophrys aveyronensis × Ophrys castellana; Ophrys ×ayusoi C.E. Hermos. & Soca
Pollinated by Andrena hattorfiana

— Ophrys aveyronensis × Ophrys ficalhoana; Ophrys ×caballeroi C.E. Hermos.
Pollinated by Andrena hattorfiana

— Ophrys aveyronensis × Ophrys passionis; Ophrys ×costei H. van Looken
Pollinated by Andrena hattorfiana

— Ophrys aveyronensis × Ophrys scolopax; Ophrys ×bernardii H. van Looken
Pollinated by Andrena hattorfiana

— Ophrys aveyronensis ×Ophrys sphegodes; Ophrys ×ezcaraiensis C.E. Hermos. & Soca
Pollinated by Andrena hattorfiana

— Ophrys balearica × Ophrys dyris
Pollinated by Megachile sicula

— Ophrys balearica × Ophrys fusca; Orchis ×spuria Reinhard
Pollinated by Megachile sicula

— Ophrys balearica × Ophrys incubacea
Pollinated by Megachile sicula

— Ophrys balearica × Ophrys lutea
Pollinated by Megachile sicula

— Ophrys balearica × Ophrys spectabilis

Pollinated by Megachile sicula

— Ophrys balearica × Ophrys speculum; Ophrys ×emmae G. Keller is H. Wettst.
Pollinated by Megachile sicula

— Ophrys balearica × Ophrys sphegodes
Pollinated by Megachile sicula

— Ophrys balearica × Ophrys sulcata
Pollinated by Megachile sicula

— Ophrys balearica × Ophrys tenthredinifera
Pollinated by Megachile sicula

— Ophrys bilunulata × Ophrys dyris; Ophrys ×provecta B. Ayuso & C.E. Hermos.

Pollinated by Andrena flavipes, Andrena nigroaenea

— Ophrys bilunulata × Ophrys fabrella
Pollinated by Andrena flavipes, Andrena nigroaenea

— Ophrys bilunulata × Ophrys fusca; Ophrys ×proxima C.E. Hermos, B. Ayuso & Soca
Pollinated by Andrena flavipes, Andrena nigroaenea

— Ophrys bilunulata × Ophrys lutea; Ophrys ×lucronii B. Ayuso

Pollinated by Andrena flavipes, Andrena nigroaenea

— Ophrys bilunulata × Ophrys sphegodes ? Ophrys bombyliflora × Ophrys fusca
Pollinated by Andrena flavipes, Andrena nigroaenea

— Ophrys bombyliflora × Ophrys guimaraesi (ut Ophrys bombyliflora × Ophrys tenthredinifera
subsp. guimaraesi)
Pollinated by Eucera nigrescens

— Ophrys bombyliflora × Ophrys incubacea; Ophrys ×cosana H. Baumann & Künkele
Pollinated by Eucera nigrescens

— Ophrys bombyliflora × Ophrys lutea; Ophrys ×clapensis M. Balayer
Pollinated by Eucera nigrescens

— Ophrys bombyliflora × Ophrys picta
Pollinated by Eucera nigrescens

— Ophrys bombyliflora × Ophrys scolopax; Ophrys ×olbiensis E.G. Camús

Pollinated by Eucera nigrescens

— Ophrys bombyliflora × Ophrys spectabilis; Ophrys ×melineae Sylviane & J.M. Moingeon

Pollinated by Eucera nigrescens

— Ophrys bombyliflora × Ophrys speculum; Ophrys ×fernandii Rolfe
Pollinated by Eucera nigrescens

— Ophrys bombyliflora × Ophrys sphegifera?
Pollinated by Eucera nigrescens

— Ophrys bombyliflora × Ophrys tenthredinifera; Ophrys ×humbertii Maire
Pollinated by Eucera nigrescens

— Ophrys castellana × Ophrys ficalhoana; Ophrys ×diez-santosii B. Ayuso Pollinated by Colletes cunicularius

— Ophrys castellana × Ophrys passionis
Pollinated by Colletes cunicularius

– Ophrys castellana × Ophrys picta
Pollinated by Colletes cunicularius

– Ophrys castellana × Ophrys scolopax; Ophrys ×vanlookeniana P. Delforge
Pollinated by Colletes cunicularius

– Ophrys castellana × Ophrys sphegodes; Ophrys ×delmeziana P. Delforge
Pollinated by Colletes cunicularius
– Ophrys castellana × Ophrys tenthredinifera
Pollinated by Colletes cunicularius

– Ophrys catalaunica × Ophrys incubacea; Ophrys ×montisciana J.E.Arnold
Pollinated by Megachile pyrenaica

– Ophrys catalaunica × Ophrys insectifera
Pollinated by Megachile pyrenaica

– Ophrys catalaunica × Ophrys passionis ? Ophrys catalaunica × Ophrys picta
Pollinated by Megachile pyrenaica

– Ophrys catalaunica × Ophrys scolopax; Ophrys ×olostensis O. Danesch & E. Danesch; Ophrys
×montserratensis Cadevall Pollinated by Megachile pyrenaica

– Ophrys catalaunica × Ophrys subinsectifera; Ophrys ×poisneliae Menos
Pollinated by Megachile pyrenaica

– Ophrys catalaunica × Ophrys speculum
Pollinated by Megachile pyrenaica

– Ophrys catalaunica × Ophrys sphegodes; Ophrys ×agustinii Kreutz
Pollinated by Megachile pyrenaica

– Ophrys clara? × Ophrys incubacea; Ophrys ×perceiana F.M. Vázquez & R. Lorenz ¿? (ut Ophrys fusca subsp. clara × Ophrys incubacea) Pollinated by Colletes cunicularius

– Ophrys corbariensis × Ophrys sphegodes
Pollinated by Eucera interrupta, Eucera longicornis, Eucera nigrescens

— Ophrys dianica × Ophrys lupercalis (forestieri); Ophrys ×lucentina P. Delforge
Pollinated by Andrena vulpecula

— Ophrys dyris × Ophrys fusca; Ophrys ×lenae M.R. Lowe & D. Tyteca pro sp.
Pollinated by Anthophora atroalba

— Ophrys dyris × Ophrys lupercalis; Ophrys ×brigittae H. Baumann
Pollinated by Anthophora atroalba

— Ophrys dyris × Ophrys lutea
Pollinated by Anthophora atroalba

— Ophrys dyris × Ophrys speculum; Ophrys ×breieri Wallenwein & Saad
Pollinated by Anthophora atroalba

— Ophrys ficalhoana × Ophrys incubacea; Ophrys ×juanae B. Ayuso
Pollinated by Eucera nigrescens

— Ophrys ficalhoana × Ophrys passionis; Ophrys ×bodegomii Benito, C.E. Hermos. & Soca
Pollinated by Eucera nigrescens

— Ophrys ficalhoana × Ophrys picta; Ophrys ×aranii E. Robles, M. Becerra, G. Astete & P. Barrena
Pollinated by Eucera nigrescens

— Ophrys ficalhoana × Ophrys riojana; Ophrys ×robatschii B. Ayuso
Pollinated by Eucera nigrescens

— Ophrys ficalhoana × Ophrys scolopax; Ophrys ×tabuencae Arnold, B. Ayuso, Hermos. & Soca
Pollinated by Eucera nigrescens

— Ophrys ficalhoana × Ophrys speculum; Ophrys ×martae B. Ayuso
Pollinated by Eucera nigrescens

— Ophrys ficalhoana × Ophrys sphegodes; Ophrys ×arizaletae Alejandre, B. Ayuso; C.E. Hermosilla & Soca
Pollinated by Eucera nigrescens

— Ophrys ficalhoana × Ophrys tenthredinifera
Pollinated by Eucera nigrescens

– Ophrys fusca × Ophrys incubacea; Ophrys ×corinthiaca Hausskn.; Ophrys ×braun-blanquetiana Soó.
Pollinated by Andrena flavipes, Andrena nigroaenea

– Ophrys fusca × Ophrys lutea; Ophrys ×gauthieri Lièvre
Pollinated by Andrena flavipes, Andrena nigroaenea

– Ophrys fusca × Ophrys passionis
Pollinated by Andrena flavipes, Andrena nigroaenea

– Ophrys fusca × Ophrys speculum; Ophrys ×fuscospeculum G. Keller
Pollinated by Andrena flavipes, Andrena nigroaenea

– Ophrys fusca × Ophrys sphegodes; Ophrys ×pseudofusca Albert & E.G. Camús
Pollinated by Andrena flavipes, Andrena nigroaenea

– Ophrys fusca × Ophrys tenthredinifera; Ophrys ×lievreae Maire
Pollinated by Andrena flavipes, Andrena nigroaenea

– Ophrys fusca × Ophrys vernixia; Ophrys ×conimbricensis O. Danesch & E. Danesch
Pollinated by Andrena flavipes, Andrena nigroaenea

– Ophrys incubacea × Ophrys insectifera; Ophrys ×alejandrei B. Ayuso
Pollinated by Andrena morio, Melecta albifrons

– Ophrys incubacea × Ophrys lutea; Ophrys ×piscinica P. & C. Delforge
Pollinated by Andrena morio, Melecta albifrons

– Ophrys incubacea × Ophrys picta; Ophrys ×maimonensis F.M. Vázquez
Pollinated by Andrena morio, Melecta albifrons

– Ophrys incubacea × Ophrys scolopax; Ophrys ×breviappendiculata Duffort; Ophrys ×llenasii
Sennen
Pollinated by Andrena morio, Melecta albifrons

– Ophrys incubacea × Ophrys speculum; Ophrys ×pantaliciensis R. Kholmüller; Ophrys ×neokelleri Soó

— Ophrys incubacea × Ophrys¿sphegifera?
Pollinated by Andrena morio, Melecta albifrons

— Ophrys incubacea × Ophrys sphegodes; Ophrys ×todaroana Macch.
Pollinated by Andrena morio, Melecta albifrons

— Ophrys incubacea × Ophrys tenthredinifera
Pollinated by Andrena morio, Melecta albifrons

— Ophrys insectifera × Ophrys lupercalis

— Ophrys insectifera × Ophrys passionis; Ophrys ×fonsaudiensis Soca

— Ophrys insectifera × Ophrys sphegodes; Ophrys ×hybrida Pokorny ex Rchb. fil.; Ophrys ×gibbosa Beck; Ophrys ×pokornyi Guetrot; Ophrys ×reichenbachiana Schulze; Ophrys ×reichenbachiana Schulze; Ophrys ×pokornyi Guetrot; Ophrys ×reichenbachiana Schulze. ×zimmermanniana A. Fuchs

— Ophrys insectifera × Ophrys subinsectifera

— Ophrys lucentina × Ophrys lutea; Ophrys ×serrae B. Ayuso

— Ophrys lucentina × Ophrys speculum; Ophrys ×marinaltae M.R. Lowe, Piera & M.B. Crespo (ut
Ophrys dianica × Ophrys speculum)

— Ophrys lucentina × Ophrys tenthredinifera; Ophrys ×donatae Tejedor, Catalá & Sospedra (ut
Ophrys dianica × Ophrys tenthredinifera)

— Ophrys lupercalis × Ophrys lutea; Ophrys ×fraresiana M.R. Lowe, Piera & M.B. Crespo

— Ophrys lupercalis × Ophrys passionis; Ophrys ×sancti-cyrensis Soca (ut Ophrys forestieri × Ophrys passionis)

— Ophrys lupercalis × Ophrys speculum; Ophrys ×eliasii Sennen (ut Ophrys forestieri × Ophrys speculum) — Ophrys lupercalis × Ophrys tenthredinifera;

— Ophrys lutea × Ophrys passionis; Ophrys ×mirandana C.E. Hermos. & G. Ubieto

— Ophrys lutea × Ophrys picta

— Ophrys lutea × Ophrys riojana; Ophrys ×fontechensis C.E. Hermos.

— Ophrys lutea × Ophrys scolopax; Ophrys ×pseudospeculum DC.

— Ophrys lutea × Ophrys speculum; Ophrys ×chobautii G. Keller

— Ophrys lutea × Ophrys sphegifera?

— Ophrys lutea × Ophrys sphegodes; Ophrys ×quadriloba E.G. Camús, Bergon & A. Camús;
Ophrys ×balarucensis G. Keller

— Ophrys lutea × Ophrys tenthredinifera; Ophry ×anamariae E. Robles & M. Becerra

— Ophrys occidentalis × Ophrys lupercalis (ut Ophrys exaltata subsp. marzuola × Ophrys lupercalis)

— Ophrys occidentalis × Ophrys tenthredinifera (ut Ophrys exaltata subsp. marzuola × Ophrys tenthredinifera (ut Ophrys exaltata subsp. marzuola × Ophrys tenthredinifera) tenthredinifera)

— Ophrys passionis × Ophrys riojana; Ophrys ×acina C.E. Hermos.

— Ophrys passionis × Ophrys scolopax; Ophrys ×hermosillae Soca & B. Ayuso

— Ophrys passionis × Ophrys speculum; Ophrys ×armentariae Ferrández, B. Ayuso & Hermos...;
Ophrys ×sanctae-cruccis A. & M. Pinaud, C.& G. Lamaurt

— Ophrys passionis × Ophrys sphegodes; Ophrys ×zamba C.E. Hermos.

— Ophrys passionis × Ophrys tenthredinifera

— Ophrys passionis × Ophrys vasconica

— Ophrys passionis × Ophrys vetula?

— Ophrys picta × Ophrys riojana; Ophrys ×zubiae C.E. Hermos. & Tabuenca

— Ophrys picta × Ophrys scolopax

— Ophrys picta × Ophrys speculum; Ophrys ×kelleriella Denis?

— Ophrys picta × Ophrys sphegodes

— Ophrys picta × Ophrys tenthredinifera; Ophrys ×gomezii E. Robles & M. Becerra

— Ophrys querciphila × Ophrys sphegodes; Ophrys ×ibanezii E. Robles, M. Becerra & A. Becerra

— Ophrys riojana × Ophrys speculum

— Ophrys riojana × Ophrys sphegodes; Ophrys ×abdita C.E. Hermos.

– Ophrys scolopax × Ophrys speculum; Ophrys ×castroviejoi Serra & Soler

– Ophrys scolopax × Ophrys sphegodes; Ophrys ×nouletii E.G. Camús; Ophrys ×philippii Gren.

– Ophrys scolopax × Ophrys tenthredinifera; Ophrys ×peltieri Maire

– Ophrys spectabilis × Ophrys tenthredinifera

– Ophrys speculum × Ophrys sphegifera ?

– Ophrys speculum × Ophrys sphegodes; Ophrys ×macchiatii E.G. Camús

– Ophrys speculum × Ophrys tenthredinifera; Ophrys ×heraultii G. Keller

– Ophrys speculum × Ophrys vasconica ?

– Ophrys speculum × Ophrys vernixia; Ophrys ×innominata D. Tyteca & B. Tyteca (ut Ophrys
lusitanica × Ophrys speculum)

– Ophrys sphegifera? × Ophrys tenthredinifera

– Ophrys sphegodes × Ophrys tenthredinifera; Ophrys ×tinyusi B. Ayuso

– Ophrys sphegodes × Ophrys vernixia ? (ut Ophrys lusitanica × Ophrys sphegodes)

INTERGENERIC NOTOTAXONS

– Cephalanthera longifolia × Orchis langei; ×Cephalorchis sussana F.M. Vázquez

– Gymnadenia conopsea × Dactylorhiza elata; ×Dactylodenia jeanjeanii G. Keller

– Gymnadenia conopsea × Dactylorhiza ericetorum ?

– Gymnadenia conopsea × Dactylorhiza fuchsii; ×Dactylodenia saint-quintinii (Godfery) J. Duvign.; ×Dactylodenia heinzeliana (Reichardt) Garay & H.R. Sweet.

– Gymnadenia conopsea × Dactylorhiza maculata; ×Dactylodenia legrandiana (E.G. Camús) E.
Peitz

– Gymnadenia conopsea × Dactylorhiza majalis; ×Dactylodenia lebrunii (E.G. Camús) E. Peit –.
Gymnadenia conopsea × Orchis morio; ×Orchigymnadenia reserata (Pau) Soó

– Gymnadenia densiflora × Dactylorhiza elata

— Dactylorhiza elata × Orchis palustris; ×Dactylocamptis caballeroi Rivas Goday (ut Orchis sesquipedalis × Orchis palustris)

— Dactylorhiza elata × Spiranthes aestivalis; ×Spilorhiza diversiflora C.E. Hermos., J. Fernández & Undagoitia

— Dactylorhiza incarnata × Orchis mascula ?

— Dactylorhiza incarnata × Serapias lingua ?

— Dactylorhiza insularis × Orchis mascula; ×Orchidactylorhiza atacina (P. Delforge) P. Delforge

— Dactylorhiza maculata × Orchis morio; ×Dactylocamptis timbaliana E.G. Camús (ut Dactylorhiza maculata × Anacamptis morio)

— Dactylorhiza maculata × Pseudorchis albida

— Dactylorhiza sambucina × Orchis fragrans; ×Dactylocamptis carpetana Willk. willk. ?

— Aceras anthropophorum × Orchis italica; ×Orchiaceras bivonae Tod.; ×Orchiaceras henriquesea (J.A. Guim.) E.G. Camús, Bergon & A. Camús; ×Orchiaceras welwitschii (Rchb. fil.) E.G. Camús

— Aceras anthropophorum × Orchis militaris; ×Orchiaceras spurium (Rchb. fil) E.G. Camús; ×Orchiaceras weddellii G. Camús

— Aceras anthropophorum × Orchis purpurea; ×Orchiaceras macra (Lindl.) E.G. Camús; ×Orchiaceras meilsheimeri (Rouy) P. Fourn.; ×Orchiaceras delamainii Alleiz. & Delamain

— Aceras anthropophorum × Orchis simia; ×Orchiaceras bergonii De Nanteuil

— Orchis coriophora × Anacamptis pyramidalis

— Orchis coriophora × Serapias lingua; ×Orchiserapias duffortii E.G. Camús

— Orchis fragrans × Anacamptis pyramidalis; ×Anacamptorchis simorrensis E.G. Camús, Bergon & A. Camús

— Orchis fragrans × Serapias parviflora; ×Orchiserapias andaluciana B. Baumann & H. Baumann

— Orchis laxiflora × Anacamptis pyramidalis; ×Anacamptorchis klingei P. Fourn...; ×Anacamptorchis larzacensis H. Kurze & O. Kurze

— Orchis laxiflora × Serapias cordigera; ×Orchiserapias nouletii (Rouy) E.G. Camús

— Orchis laxiflora × Serapias lingua; ×Orchiserapias complicata E.G. Camús

— Orchis morio × Serapias cordigera; ×Orchiserapias monfortensis De la Peña

— Orchis morio × Serapias lingua; ×Orchiserapias capitata De Laramb. ex E.G. Camús; ×Orchiserapies jeanjeanie G. Keller

— Orchis morio × Serapias vomeracea; ×Orchiserapias leroyi E.G. Camús & Sennen; ×Herorchiserapias fontanae (Rigo & Goiran) P. Delforge

— Orchis papilionacea × Anacamptis pyramidalis ¿?; ×Anacamptorchis vanlokeinii C. Bernard & G. Fabre

— Orchis papilionacea × Serapias lingua ?, ×Serapimeulenia barlae (Barla ex K. Richter) P. Delforge

— Orchis papilionacea × Serapias cordigera ?, ×Orchiserapias debeauxii E.G. Camús

— Himantoglossum hircinum × Orchis simia; ×Orchimantoglossum lacazei (E.G. Camús) Asch. & Graebn.

— Barlia robertiana × Orchis collina; ×Barliorchis almeriensis Pallarés

— Anacamptis pyramidalis × Serapias lingua; ×Serapicamptis forbesii Godfery

BIBLIOGRAPHY .

Dressler, R.L., 1976. How to study orchid pollination without any orchids. In Senghas, K., ed.
Proceedings 8th World Orchid Conference. Frankfurt: Eighth World Orchid Conference, 534-537.

Nilsson, L.A. 1983. Processes of isolation and introgressive interplay between *Platanthera bifolia* (L.) Rich and *P. chlorantha* (Custer) Reichb. (Orchidaceae). *Botanical Journal of the Linnean Society,* 87: 325-350.

Peakall, R. 1990. Responses of male *Zaspilothynnus trilobatus* Turner wasps to females and the sexually deceptive orchid it pollinates. *Functional Ecology,* 4: 159-167.

Scaccabarozzi, D., Cozzolino, S., Guzzetti, L., Galimberti, A., Milne, L., Dixon, K.W. & Phillips, R.D. 2018. Masquerading as pea plants: behavioural and morphological evidence for mimicry of multiple models in an Australian orchid. *Annals of Botany,* 122: 1061-1073.

Vereecken, N.J., Wilson, C.A., Hötling, S., Schulz, S., Banketov, S.A. & Mardulyn. P. 2012. Pre- adaptations and the evolution of pollination by sexual deception: Cope's rule of specialisation revisited. *Proceedings of the Royal Society B: Biological Sciences,* 279: 4786-4794.

MIX
Papier aus verantwortungsvollen Quellen
Paper from responsible sources
FSC® C105338

Printed by Books on Demand GmbH, Norderstedt / Germany